A Compendium of Good Practices in Biotechnology

BOOKS IN THE BIOTOL SERIES

The Molecular Fabric of Cells
Infrastructure and Activities of Cells

Techniques used in Bioproduct Analysis
Analysis of Amino Acids, Proteins and Nucleic Acids

Principles of Cell Energetics
Energy Sources for Cells
Biosynthesis and the Integration of Cell Metabolism

Genome Management in Prokaryotes
Genome Management in Eukaryotes

Crop Physiology
Crop Productivity

Functional Physiology
Cellular Interactions and Immunobiology
Defence Mechanisms

Bioprocess Technology: Modelling and Transport Phenomena
Operational Modes of Bioreactors

In vitro Cultivation of Micro-organisms
In vitro Cultivation of Plant Cells
In vitro Cultivation of Animal Cells

Bioreactor Design and Product Yield
Product Recovery in Bioprocess Technology

Techniques for Engineering Genes
Strategies for Engineering Organisms

Principles of Enzymology for Technological Applications
Technological Applications of Biocatalysts
Technological Applications of Immunochemicals

Biotechnological Innovations in Health Care

Biotechnological Innovations in Crop Improvement
Biotechnological Innovations in Animal Productivity

Biotechnological Innovations in Energy and Environmental Management

Biotechnological Innovations in Chemical Synthesis

Biotechnological Innovations in Food Processing

A Compendium of Good Practices in Biotechnology

BIOTECHNOLOGY BY OPEN LEARNING

A Compendium of Good Practices in Biotechnology

PUBLISHED ON BEHALF OF :

Open universiteit   and   **University of Greenwich (formerly Thames Polytechnic)**

Valkenburgerweg 167
6401 DL Heerlen
Nederland

Avery Hill Road
Eltham, London SE9 2HB
United Kingdom

Butterworth-Heinemann

The Biotol Project

The BIOTOL team

OPEN UNIVERSITEIT, THE NETHERLANDS
Prof M. C. E. van Dam-Mieras
Prof W. H. de Jeu
Prof J. de Vries

UNIVERSITY OF GREENWICH (FORMERLY THAMES POLYTECHNIC), UK
Prof B. R. Currell
Dr J. W. James
Dr C. K. Leach
Mr R. A. Patmore

This series of books has been developed through a collaboration between the Open universiteit of the Netherlands and University of Greenwich (formerly Thames Polytechnic) to provide a whole library of advanced level flexible learning materials including books, computer and video programmes. The series will be of particular value to those working in the chemical, pharmaceutical, health care, food and drinks, agriculture, and environmental, manufacturing and service industries. These industries will be increasingly faced with training problems as the use of biologically based techniques replaces or enhances chemical ones or indeed allows the development of products previously impossible.

The BIOTOL books may be studied privately, but specifically they provide a cost-effective major resource for in-house company training and are the basis for a wider range of courses (open, distance or traditional) from universities which, with practical and tutorial support, lead to recognised qualifications. There is a developing network of institutions throughout Europe to offer tutorial and practical support and courses based on BIOTOL both for those newly entering the field of biotechnology and for graduates looking for more advanced training. BIOTOL is for any one wishing to know about and use the principles and techniques of modern biotechnology whether they are technicians needing further education, new graduates wishing to extend their knowledge, mature staff faced with changing work or a new career, managers unfamiliar with the new technology or those returning to work after a career break.

Our learning texts, written in an informal and friendly style, embody the best characteristics of both open and distance learning to provide a flexible resource for individuals, training organisations, polytechnics and universities, and professional bodies. The content of each book has been carefully worked out between teachers and industry to lead students through a programme of work so that they may achieve clearly stated learning objectives. There are activities and exercises throughout the books, and self assessment questions that allow students to check their own progress and receive any necessary remedial help.

The books, within the series, are modular allowing students to select their own entry point depending on their knowledge and previous experience. These texts therefore remove the necessity for students to attend institution based lectures at specific times and places, bringing a new freedom to study their chosen subject at the time they need and a pace and place to suit them. This same freedom is highly beneficial to industry since staff can receive training without spending significant periods away from the workplace attending lectures and courses, and without altering work patterns.

SOFTWARE IN THE BIOTOL SERIES

BIOcalm interactive computer programmes provide experience in decision making in many of the techniques used in Biotechnology. They simulate the practical problems and decisions that need to be addressed in planning, setting up and carrying out research or development experiments and production processes. Each programme has an extensive library including basic concepts, experimental techniques, data and units. Also included with each programme are the relevant BIOTOL books which cover the necessary theoretical background.

The programmes and supporting BIOTOL books are listed below.

Isolation and Growth of Micro-organisms
Book: *In vitro* Cultivation of Micro-organisms
          Energy Sources for Cells

Elucidation and Manipulation of Metabolic Pathways
Books: *In vitro* Cultivation of Micro-organisms
          Energy Sources for Cells

Gene Isolation and Characterisation
Books: Techniques for Engineering Genes
          Strategies for Engineering Organisms

Applications of Genetic Manipulation
Books: Techniques for Engineering Genes
          Strategies for Engineering Organisms

Extraction, Purification and Characterisation of an Enzyme
Books: Analysis of Amino Acids, Proteins and Nucleic Acids
          Techniques used in Bioproduct Analysis

Enzyme Engineering
Books: Principles of Enzymology for Technological Applications
          Molecular Fabric of Cells

Bioprocess Technology
Books: Bioreactor Design and Product Yield
          Product Recovery in Bioprocess Technology
          Bioprocess Technology: Modelling and Transport Phenomena
          Operational Modes of Bioreactors

Further information: Greenwich University Press,
University of Greenwich, Avery Hill Road, London SE9 2HB.

Contributors

AUTHORS

C. Dadomo, University of the West of England, Bristol, UK.

Professor M.C.E. van Dam-Mieras, Open universiteit of the Netherlands, Heerlen, The Netherlands.

Dr. D. Dickerson, University of the West of England, Bristol, UK

Dr. H.J.M. van de Donk, National Institute of Public Health and Environmental Protection, Bilthoven, The Netherlands.

Dr. J. Greenman, University of the West of England, Bristol, UK.

Dr. C. K. Leach, De Montfort University, Leicester, UK.

T.W. Roberts Zeneca Agrochemical, Bracknell, Berkshire, UK.

Dr P. S. Scull, University of the West of England, Bristol, UK.

Dr. D. Shaw, University of the West of England, Bristol, UK.

Dr. G. Tuijneuburg Muijs, Prins Bernhardkade, Rotterdam, The Netherlands.

EDITORIAL TEAM

Dr. J. Greenman, University of the West of England, Bristol

Dr. C. K. Leach, De Montfort University, Leicester, UK.

Prof M. C. E. van Dam-Mieras, Open universiteit, Heerlen, The Netherlands

SCIENTIFIC AND COURSE ADVISORS

Dr. C. K. Leach, De Montfort University, Leicester, UK.

Prof M. C. E. van Dam-Mieras, Open universiteit, Heerlen, The Netherlands

ACKNOWLEDGEMENTS

Grateful thanks are extended, not only to the authors, editors and course advisors, but to all those who have contributed to the development and production of this book. They include Miss K Brown and Mrs S. Smith .

We are particularly grateful to the Office for Official Publications of the European Communities for permission to reproduce articles from the Official Journal.

The development of this BIOTOL text has been funded by **COMETT, The European Community Action Programme for Education and Training for Technology**. Additional support was received from the Open universiteit of The Netherlands and by University of Greenwich (formerly Thames Polytechnic).

Contents

Preface

The BIOTOL series of texts predominantly focus on the scientific and technical principles and procedures applicable to biotechnology. Biotechnological activities are, however, greatly influenced by an extensive range of regulatory requirements and recommendations designed to protect investors, inventors workers, consumers and the environment. It is essential therefore that those who practice within biotechnology are aware of, and respond to, these regulatory requirements. Individual BIOTOL texts, refer to and, to varying extents, discuss these requirements in the context of the specific topics under consideration. This text greatly augments these discussions by bringing together a wide spectrum of regulatory affairs relating to biotechnological activities. It does not aim to provide detailed copies of all of the relevant national and supra-national legislation since this would have resulted in the production of a major encyclopedia. Indeed by providing a digest of the issues and regulations, this text allows the reader to identify the relevant regulations from accessible sources and to become familiar with their principles and application. The text is, therefore, designed to provide an easy to use resource of regulatory affairs and good practices for students and researchers engaged in biotechnological pursuits.

Within Europe, most, if not all, regulations that affect biotechnology (eg contained use of genetically manipulated organisms, deliberate release of genetically modified organisms, authorisation of medicinal products for human and veterinary use, laws that concern protection of workers and the environment etc) emanate from the European Community (EC). Similarly the EC has an important role in stimulating and resourcing biotechnological enterprise. Many biotechnologists have only a fragmentary knowledge of the activities of the EC and even less understanding of the structure and roles played by the various EC institutions in the regulation and development of biotechnology. Moreover, most students (and many researchers) do not know how to obtain copies of the relevant regulations or other EC publications of importance. This text begins, by briefly examining the formation and administrative organisation of the EC and explains the different forms (regulations, decisions and directives) of EC measures. The text also explain the principle sources of these measures and explains through the use of specific examples, the general structure and format of EC legislation.

The second chapter provides a contextual overview of the legislation of direct relevance to biotechnology. The bulk of the text, however, deals with legal and operation issues which have generic application in biotechnology. These include the principles and practices of Good Laboratory and Good Manufacturing Practices (GLP, GMP), the safe use of micro-organisms and ionising radiation, risk evaluation and management of operations and processes involving genetic manipulation and the use of animals in the laboratory. The issues of intellectual property rights, including the application of patenting procedures and breeders rights, are also explained.

Towards the end of the text, special consideration is given to the legislative framework applicable to the introduction of biotechnological products and process in health care, food processing and the production of fine chemicals. Of special importance in these major business sectors is the fulfilment of the regulatory requirements necessary to gain market authorisation and key to this authorisation is the need to compile a suitable Market Authorisation Dossier (file). The purpose, content and processing of such dossiers are described.

The author: editor team have brought a wealth of experience both as contributors to biotechnology research and development and as trainers of biotechnologists. The outcome of their efforts is a resource which will be valued highly by students and their teachers as well as those engaged in research and development.

Scientific and Course Advisors: Professor MCE van Dam-Mieras
Dr. C K Leach

EC structure and Tools

EC structure and Tools

1.1 Introduction

The relationships between the European Community (EC) and the development of Biotechnology is important for two main reasons. Firstly, the EC is a source of resources such as research grants and development funds. A considerable amount of these are directed towards new innovation and technology and this certainly includes biotechnology. Secondly, the EC has laws, directives, regulations and guidelines, many of which affect directly or indirectly the development and use of biotechnology. It is the purpose of the following chapters to introduce the reader to the structure and function of the EC and attempt to remove the mystery which surrounds this subject as it affects biotechnology.

We will begin by briefly describing the foundation of European institutions. We will then examine the insititutions that formulate, update, modify and implement and monitor EC policies. In the final part of this chapter, we will outline the role of the EC in the development of Biotechnology both as a source of funding and as a regulator. The aim of the first chapter is therefore to provide a context in which the more specific issues of later chapters can be explained and discussed.

1.2 The foundations of Europe

Following the second world war, the countries of western Europe have steadily increased their ties and trading links with each other. Countries have learnt to cooperate with their European neighbours for the common good. The original aim of this cooperation was to maintain peace and prevent further war. However, an additional bonus of the closer working links has been to aid economic growth of both individual European Countries and Europe as a whole.

The EC effectively began in 1952 when Italy, France, Belgium, the Netherlands, Luxembourg and West Germany formed the European Coal and Steel Community in order to encourage reconciliation and trade after the second world war. In 1958, the

EEC

same countries formed the European Economic Community (EEC) whose purpose was to strengthen economic ties and create a common market. This was based on an agreement which was signed on March 25, 1957 in Rome and has become known since

Treaty of Rome

as 'The Treaty of Rome'. This important treaty meant that member states could trade freely with each other and encourage mutual cooperation to help develop trade, agriculture and industry. It was hoped that the treaty would help to raise living standards and ensure that war between member states would become "impossible". However, the formation of the EEC had little effect on the daily life of the average European citizen and it did not become directly involved in national domestic policies such as defence, foreign policy, health and safety, education, etc.

EC

In 1967, the original EEC countries became known as the European Community or the EC. The United Kingdom had not originally wanted to join the EEC since it wished to maintain its links with the Commonwealth and what remained of the British Empire. However, it did attempt to join the EEC/EC in 1961 and again in 1967 only to be refused membership by one of the member states (France, under President Charles de Gaulle) on the grounds that the UK was considered to be too close an ally of the United States to commit itself whole-heartedly to Europe. However, in 1972 the UK (under Prime Minister Edward Heath) did sign the First Treaty of Accession which signified the admission of four new members (Ireland, Denmark, Norway and the UK) to the European Community. Norway did not in the end join due to a negative result obtained from running a national referendum on the subject. The Treaty consisted of articles stating that the four new member states accede to the existing Communities and accept all their rules. Full membership was conditional upon the incorporation of Community Law into the municipal laws of the new member states. The period of transition was completed by July 1977, this being the date that the new community of nine was established. In 1975 a referendum was held in the UK under the Labour Government of Harold Wilson. In this, 67 % of voters showed that they wished the UK to remain within the EC which it continued to do. Greece joined the EC in 1981 as did Spain and Portugal in 1986. Only one territory has left the EC, this being Greenland (originally part of Denmark) which left in 1985.

Single Eurpean Act 1986

The main thrusts and basic principles underlying the common will of the member states of the EC can be found from the various treaties, starting with the treaty of Rome in 1957 and culminating via successive reworkings of the treaties up until the Single European Act (SEA) in 1986. These basic principles include:

- the wish to build a single European Market, together with the wish to abolish physical borders and to prohibit non-tariff customs barriers (technical, fiscal, legal) between Member States, and to establish the free movement of goods, persons, services and capital;

- the equality of businesses with respect to the principles of free competition;

- the desire to raise the less developed regions of the Community to the average economic level of the Community;

- the desire to increase the competitiveness of European industry by means of technological progress;

- the development of employment and training;

- the conservation of the environment.

None of the basic principles are described exhaustively since common policies have to be modified in detail or new ones implemented in order to keep them in line with new developments in society or changes in the Governments of Member States.

More recently, a further strengthening of ties between European States has been developed through the 1992 Maastricht agreement. However, at the time of writing, this agreement is yet to be fully implemented.

1.3. European Institutions

Numerous European Institutions have arisen in order to formulate, update, modify, implement or monitor community policies. The most important institutions are the following:

- The Commission;

- The European Parliament;

- The Council;

- The European Court of Justice;

- Ancillary Community Bodies.

1.3.1. The Commission

The European Commission is based in Brussels, Belgium. It is the institution where most of the day-to-day affairs are handled. It is led by 17 commissioners none of them being elected but selected and appointed by a common agreement of the governments of the member states; they serve for four years. For example, the UK nominates two commissioners whilst some countries (eg. Greece) only have one. Of the 17 commissioners, one is appointed as President of the Commission and six are appointed as vice-presidents to serve in office for two years. All may be re-appointed at the end of their terms of office (e.g., at the Lisbon summit meeting in June 1992, Jacques Delors was reconducted as President of the Commission for another two-year term).

It is the role of the Commission to propose and draft (ie initiate) new laws and implement policies once they have been agreed by the Council of Ministers (and, where provided for by the treaty, after consultation with the European Parliament). Only the Commission can draft legislation, whereas only the Council of Ministers has the power to adopt (but not modify) legislation. Apart from this legislative role, the Commission has also been vested with executive powers to ensure the enforcement of decisions made by the Council. Notably, in connection with the Common Agricultural Policy, the Council has delegated wide decision-making powers to the Commission under article 155 EEC. This role was subsequently reinforced under article 145 EEC as amended by the 1986 Single European Act (SEA).

The Commission has also been granted representative as well as financial functions - under the former, the Commission represents the Community in its relationship with non-member states and other international organisations (such as the GATT where the Community negotiates *in lieu* of the member-states. GATT = General Agreement on Tariffs and Trade) Under the latter, the Commission is in charge of drafting the budget of the Community and of implementing it once it is finally adopted by the Council and the Parliament.

Finally, the Commission acts as a watch-dog to ensure that the Treaties obligations are observed by the member-states. The Commission also administers EC funds and subsidies. It is required to act in complete independence for the good of the Community.

Not surprisingly, the executive role of the Commission requires a vast apparatus of committees and civil servants (most based in Brussels, but some in Luxembourg). The majority of the civil servants are recruited on a permanent basis.

To give an idea of the complexity of the Commission's workings, there are 23 divisions called "Directorates-General" which have distributed duties (see Table 1.1 for a list of these).

Margin notes:
non-elected Commissioners

legislative role of the Commission enforcement of Council decision

Represenative role of the Commission

independence of the Commission

Directorates-General

1.	**Directorates-General**
DG I	external relations
DG II	economic and financial affairs
DG III	internal market and industrial affairs
DG IV	competition
DG V	employment, industrial relations and social affairs
DG VI	agriculture
DG VII	transport
DG VIII	development
DG IX	personnel and administration
DG X	audio-visual, information, communication and culture
DG XI	environment, nuclear safety, civil protection
DG XII	science, research, development
DG XIII	telecommunications, information technology and innovation
DG XIV	fisheries
DG XV	financial institutions and company law
DG XVI	regional policy
DG XVII	energy
DG XVIII	credit and investment
DG XIX	budgets
DG XX	financial control
DG XXI	customs union and indirect taxation
DG XXII	co-ordination of structural policies
DG XXIII	business policy, trade, tourism and social economy

2. Committees instituted directly by European Treaties

Advisory Committee of the ECSC
The Economic and Social Committee
Monetary Committee
European Social Fund Committee
Advisory Committee on Transport
Technical and Scientific Committee of EURATOM

3. Secondary Committees (set up by institutions)

European Research and Development Committee (CERD)
Scientific and Technical Committee (CST)
Advisory Committee of the Joint Research Centres (JRC)
Standing Committee on Agricultural Research
Scientific Committee for Food
Group of Specialists "Rational Use of Energy"
Economic Policy Committee
Economic Development Fund Committee
ERDF Committee
Regional Policy Committee
Advisory Committee on Safety, Hygiene and Health Protection at work
Advisory Committee on Freedom of Movement for Workers
Advisory Committee on Vocational Training
Advisory Committee for Public Works Contracts
Committee of Governors of the Central Banks

Table 1.1 Directorates-General and Committees of the European Commission.

In addition, there are other specialist services which support the Commission (or the Commission's Directorates-General). Examples include the Consumer Policy Service, the Task Force for Human Resources, Education, Training and Youth, and the EURATOM Supply Agency. Moreover there is a Business Co-operative Centre (BCC) and the Office for Official Publications of the EC. The Office for Official Publications is responsible for all the EC's publications both official and otherwise. More information on the types of publications available is given in Table 1.2 here.

1. The Official Journal

The Official Journal (or OJ as it is called) publishes regulations, directives and decisions. This appears almost daily in each of the nine official languages of the EC. There are three different series of the OJ. The 'L' series is devoted to secondary legislation; the 'C' series publishes official information and communications; the 'S' series publishes public contract notices, supply tenders, Invitations to Tender of the European Development Fund etc. (See Appendix 1.2 for address of the Official Journal in Member States

2. EC Catalogues (published by the Office for Official Publications of the EC)

a. Publications of the European Communities

This catalogue is published quarterly in all the nine official languages of the EC. It lists all the reports, studies, monographs, periodicals and series of documents produced by the EC Institutions except the documents which relate to the Commission, the European Parliament and the Economic and Social Committee. Documents which relate to these institutions are published in a catalogue which is simply called the "Documents" catalogue.

b. Documents catalogue

This catalogue is published monthly and includes all the official documents published by the Commission, the Parliament and the Economic and Social Committee (referred to as COM, PEDOC and ESC documents respectively).

3. Directory of EC Information Sources

This publication is produced annually outside the Commission by Euroconfidentiel, (B.P. 29 - B-1330 Rixensart, Belgium). It is a 900 page book which covers all the EC's areas of work in a single volume.

4. Information from the Commission

Many of the Directorates-General have their own information systems/structures and their own libraries (scattered throughout Brussels and Luxembourg). DG X (Information, communication and culture) is devoted entirely to information, communication and culture. A regular publication by DG XIII is called "Information and Technology Transfer" (ITT) and it contains information on research policy, research programmes, conferences and symposia.

5. Electronic information

a. European Community Host Organisation (ECHO)

This is based in Luxembourg and is the information service of Directorate-General XIII (Telecommunications, Information Industries and Innovation). ECHO gives free access to many of the EC's data banks (eg. general data on the EC economy and industry, information on European research and development programmes, projects and publications) and includes systems for teaching and familiarising the new user on "electronic information services" and other assistance in the use of data banks.

b. Community R & D Information Service (CORDIS)

The data banks on CORDIS allow the user to gain information on the current and past Research and Technological Development (RTD) Programmes. A detailed description of RTD-Projects supported by the RTD-Programmes is also available.

6. Information Specifically on Biotechnology

a. European Biotechnology Information Service (EBIS)

The European Biotechnology Information Service (EBIS) publishes a newsletter called "CUBE". This gives information about all Community matters relating to biotechnology.

b. European Biotechnology Newsletter (EBN)

This non-Community publication is provided by an independent group and provides up-to-date concise information on European biotechnology, particularly from a scientific, industrial and legislative viewpoint. Subscription is available from: EBN - European Biotech Newsletter, ELSEVIER Publications, 29, Rue Buffon, F-75005, Paris.

Table 1.2 EC Information and Publications.

To summarise; the Commission is the main initiator of Community policies within the broad framework defined by the various Treaties. The role of the Commission is major since it holds the real Community executive power. This revolves around five main responsibilities:

- the drafting of texts in application of the provisions of the Treaties;

- the implementation of policies to specific cases, eg for agriculture, for competition, the environment, etc.;

- the management of Community funds, eg. the European Social Fund (ESF), the European Regional Development Fund (ERDF), Euratom funds, etc.;

- the management of safeguard clauses in the treaties, for derogations to the main principles defined therein;

- the noting of infringements and prosecution.

Thus, it is the Commission that will establish infringements of Community decisions by Member States. If necessary, the Commission can refer a matter to the Court of Justice, whose judgements are binding both for individuals and Member States and the Community institutions.

1.3.2 The European Parliament

MEPs

The Parliament sits in Strasbourg, France. It has 518 representatives (MEPs) including 81 from each of France, Germany, Italy and the UK. The other EC countries send a smaller number. Representatives are elected every five years.

minutes published in the Official Journal

The European Parliament is the only EC institution to have public debates. All the minutes of the sessions are published in the Official Journal (OJ). As in national assemblies, the delegates tend to sit in their political groupings (transnational) during sessions. For EC parliamentary purposes a political group must have a minimal number of delegates and must be represented in committees and sessions by a spokesperson.

parliamentary bureau

The activities of the Parliament are the responsibility of the Parliamentary Bureau which is composed of a President and 14 Vice-Presidents who are elected by Parliament to serve for two and a half years. The agenda for the plenary sessions is prepared by the Bureau plus the Presidents of the various political groups. Five people called "quaestors" are also elected by parliament to carry out administrative duties and act in an advisory capacity at the Bureau.

There are at the present time 18 specialist permanent committees which prepare the work of the plenary sessions. They also function in following up proceedings and act as mediators between the political groups. The appropriate specialist committee may be called by Parliament for public consultation if technical questions arise during sessions.

legislature powers of the Parliament increased by Single European Act and the Treaty of Maastricht

The Parliament scrutinises the Commissions proposals for legislation and votes on amendments. It can only hold "Opinions" on proposals or amended proposals. However, it is unlikely that Council would take any major decision without the support of Parliament. The legislative powers of the Parliament were first strengthened by the cooperation procedure, introduced by the Single European Act, which gives the Parliament an opportunity to propose amendments to draft legislation; and secondly by a co-decision procedure introduced by the Treaty of Maastricht on Political Union (not yet into force since the ratification process is not completed), which gives the Parliament a right of veto in matters subject to this new procedure. It also decides, jointly with the Council, the Community budget. It also has a general supervisory role over the activities of the EC institutions and it can censure the Commission. Although its powers are clearly defined, its scope is very wide indeed since it can address any problem facing the Community. Thus the Parliament may consider any matter which may affect the lives of European citizens from human rights to genetic manipulation.

1.3.2.The Council

The Council of the European Community (more usually referred to as "the Council") is the law-making body of the EC. The Council Presidency is responsible for convoking the Council and chairing its meetings and is held in a given order by each of the Member States for a period of six months (from 1993-1998 the order is Denmark, Belgium, Greece, Germany, France, Spain, Italy, Ireland, the Netherlands, Luxembourg, UK and Portugal). With regard to the rest of the members, the Council represents, at EC level, the governments in power in each Member State which delegates some of its representatives to serve on the Council. Since each Member State may select different officials as delegates depending on the topic under consideration, and since Governments themselves are not permanent, it can be seen that the actual composition of the Council with respect to individuals is not fixed but is always changing. What remains fixed is the ratio of votes that are accorded to each Member State. Thus 10 votes are given to each of France, Germany, Italy and the UK; 8 for Spain, 5 each for Belgium, Greece, the Netherlands and Portugal, 3 for Denmark and 2 for Luxembourg. This distribution favours the smaller countries since they have more votes for their size than would be justified by comparing the proportions of their populations.

Changing composition of the council

A number of bodies assist the Council.

The General Secretariat of the Council

This body consists of community officials and helps both national civil servants and representatives at the Council to carry out their work. The general secretariat is responsible to the Secretary General and is divided up into sections (Directorates-General) each of which specialises in a specific area.

A number of Directorates-General and committees are in existence which, like those attached to the European Commission, help support the Council or represent outside groups within the Council. (Table 1.3 gives some examples of these).

1.	**Directorates-General**
DG A	Personnel, administration, information, publications and documentation
DG B	Agriculture and fisheries
DG C	Internal market, industrial policy, approximation of legislation, free circulation of services and intellectual property
DG D	Research, Energy, Environment, Transport, Consumers
DG E	External relations and development
DG G	Economic, financial and social affairs
2.	**Committees**
	Permanent Representatives Committee (COREPER)
	Committee for Scientific and Technical Research (CREST)
	Committee for Scientific and Technical Co-operation (COST)
	Article 113 Committee
	Special Committee on Agriculture
	Energy Committee
	Standing Committee on Employment
	Education Committee
	Select Committee on Co-operation Agreements
	Standing Committee on Uranium Enrichment (COPENUR)

Table 1.3 Directorates-General and Committees attached to the Council.

The Conference of Representatives of the Governments of Member States

This body plays a part in decision-making in certain specific areas laid down in the European Treaties and also plays a role in modifying or adopting new treaties.

The European Council

The European Council (not to be confused with the Council of the European Community nor the Council of Europe) was first formed in 1974 and is composed of the Heads of State and Heads of Government of the Member States. These meetings are held twice (or three times) a year and are chaired by the country presiding over the Council. The European Council defines the common policies to the Member States and proposes policy to the Council. The European Council has important influence and can give impetus to the progressive development of EC policies on many different levels (eg. monetary, economic, legislative, social).

In short, the Council of the European Community is the law-making body of the EC. Most measures are decided by majority vote except when amendments to the Commissions proposals are to be adopted when agreement has to be unanimous (excluding abstentions). Unanimity is also required for other important decisions such as agreement on changes to Value Added Tax within the community.

In general the Council can only act on the Commission's initiative, by adopting a proposal put forward by the Commission, and examined by the Parliament and the ESC (see Figure 1.1).

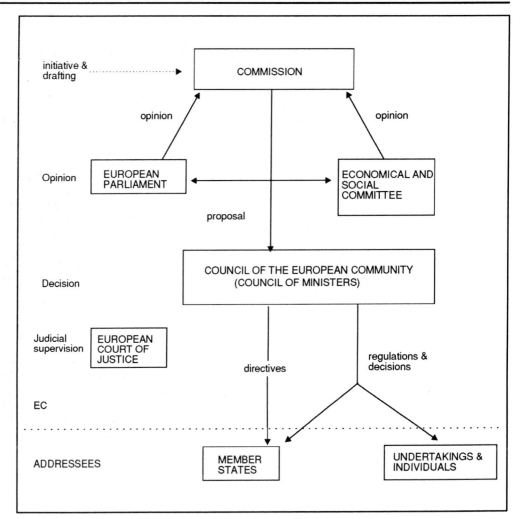

Figure 1.1 Diagram of the EC decision-making process (simplified).

The Council's decisions may take different forms which (in increasing order of importance) are: opinions, recommendations, decisions, Directives and Regulations. (These are referred to in Table 1.4.).

Regulations

Have a binding general effect for all the Member States and are applicable immediately after their publication in the OJ. The Regulation prevails over any texts of law which might be in force under the national laws of the Member States. Its application is upheld by the European Court of Justice before which any European legal entity can seek redress, even against a Member State. It is the Council that makes regulations, except in cases of delegated powers to the Commission (e.g. in the field of agricultural policy).

Directives

Give national authorities the power to legislate, as long as they respect the minimum objectives laid down in the Directive. Directives take effect as soon as they are published in the OJ. They define the lower limits above which the national laws of Member States must be in order to respect the Community position. Following the publication of the Directive, the Member State is obliged to adapt its national laws within a given time limit. A State can be sentenced by the Court of Justice for not incorporating a Directive into its own law. However, in principle, a Directive cannot itself form the basis for a sentence. (ie. whereas regulations are binding in their entirety, directives are binding as to the result to be achieved, upon each Member State to which they are addressed). It is the Council that makes Directives.

Decisions

Are mandatory for those to whom they are addressed which could include Member States, corporations, or individuals. Decisions can be made by the Council or by the Commission and like a regulation or a directive has to be substantiated (ie based on a Treaty provision). Unlike a regulation, a decision is binding upon the addressee only. Unlike a directive, a decision is "binding in its entirety" which leaves no discretion as to the manner in which it is to be carried out.

Recommendations

Are not binding and therefore represent "suggestions" or "advice" which may or may not be followed by the addressee.

Opinions

Are not binding and therefore represent "a point of view" reached by the Council, Commission, Parliament or the ESC.

Table 1.4   Regulations, Directives, Decisions, Opinions and Recommendations.

1.3.3 The European Court of Justice

ECSC agreement The European Court was created at the time of the very first European Treaty, the European Coal and Steel Community (ECSC) agreement in 1952. It has remained the legal institution of the EC since that time and its role is to ensure that both primary laws contained in the various European Treaties, and all the secondary laws are respected. It sits in Luxembourg and has 13 judges appointed by the Member States. The 13 judges co-opt their president from amongst themselves. The judges are assisted by six advocates-general who are also chosen by the Member States. Both judges and advocates-general serve for six years and are chosen from amongst the best legal experts available and should therefore be of the highest independence and competence.

The European Court can quash any measure of the institutions and interpret Community law. The institutions, as Member States and to a lesser extent, undertakings and individuals, can bring cases to the court. Its decisions are binding on the Member States.

1.3.4. Ancillary Community Bodies

There are a number of ancillary institutions and bodies which participate in the activities of the Communities or advise. They have their origin either in the treaties or have been established in the light of experience.

Economic and Social Committee (ESC)

The Economic and Social Committee (ESC) is the most typical example of the former category. It was instituted by the Treaties of Rome in 1957 in order to make economic and social interest groups party to the Community decision-making process. This is the only European Committee where its role has been specifically stipulated in the European Treaties and which can offer opinions on the Commissions proposals to the Council (see Figure 1.1). This committee can also issue its "own-initiative" opinions.

composition of ESC

The committee itself is composed of three groups of representatives (i) employers, (ii) workers, (iii) various interest groups (eg. trade, transport, consumers, agriculture, etc.) and has 189 members from the Member States. The committee appoints its own Chairman and Bureau (to serve for two years) from amongst themselves. The Bureau is made from 30 members and its role is to organise and manage the business of the committee. To deal with all the different areas handled by the EC, the committee is divided into nine sections. As stated previously, the committee can offer "Opinions" on Commissions proposals.

advisory role of ESC

In short, the Economic and Social Committee is an advisory body to which the Council and Commission refer in the Community decision-making process (see Figure 1.1). The European Treaties stipulate that its role should be to advise on any decisions which relate to agriculture, the free movement of workers, transport, the right of establishment, social policy, European Social Fund (ESF) and certain other subjects.

The Court of Auditors, the European Bank of Investment, the European Social Fund and various committees such as the Scientific and Technical Committee of the EURATOM, the Monetary Committee, etc., are other examples of ancillary Community bodies originating from the treaties.

Examples of the second category (those established in the light of experience) include the so-called "management" and "regulatory" committees which assist the Commission exercising its delegated powers of execution, notably, in the field of agriculture.

1.3.5 Other bodies of importance in the EC

There are many of these. Here we list some of these including those of particular relevence to Biotechnology. These are:

- the European Molecular Biology Organisation (EMBO);
- the European Molecular Biology Conference (EMBC);
- the European Patent Organisation (which is responsible for the European Patent Office which handles European patent applications).
- the European Foundation for the Improvement of Living and Working Conditions;
- the African, Caribbean and Pacific - EC Council (ACR-ECC);
- the European Organisation for Nuclear Research (CERN).

1.3.6 The Council of Europe

The Council of Europe is not to be confused with either the Council of the European Communities nor the European Council since it is both older (set up in 1949) and has a wider membership (27 countries) than the EC. Its objectives are to:

- work towards an ever-closer European Union;

- promote the principles of democracy and human rights;

- improve the standard of life and promote human values;

It consists of the following institutions:

- the Committee of Ministers (that is Ministers of Foreign Affairs from the member countries who meet twice a year);

- the Parliamentary Assembly (177 members from the member countries, meet three times a year in Strasbourg; formed in 1949);

- the General Secretariat (assists the Council in its administrative duties);

- the European Commission on Human Rights (27 independent lawyers, one from each of the member countries);

- the European Youth Foundation and Centre

1.4 Role of European Community in the development of Biotechnology

As mentioned at the beginning of this chapter, the EC is an important factor in the development of Biotechnology for two main reasons. Firstly, the EC is a source of resources such as research grants and development funds and secondly the EC has laws, directives, regulations and guidelines, many of which affect directly or indirectly the development and use of biotechnology.

1.4.1. The EC as a source of research, development and training funds

The European Community collects revenues from a number of sources including:

- a proportion of Value Added Tax from each Member State;

- revenues from customs duties;

- agricultural levies;

- payment of a proportion from Member States based on the sum of all the Member States' gross national product (at market prices defined in accordance with Community rules).

This vast amount of money is then paid out in order to cover spending on:

- the administration, buildings, resources of the various European Institutions;

- payment for contracted (tendered) services;

- loans made out to fund Community projects;

- subsidies.

EIB

With regard to loans, most of these are allocated by the European Investment Bank (EIB) which raises its own funds by borrowing on the public capital markets. Often, loans can be cumulated with subsidies granted by the Structural Fund.

ESF, ERDF
EAGGF IMPs
RTD

With regard to subsidies, these can be paid out to priority Community projects by the Structural Fund using systems or instruments managed by the Commission. These include projects funded by the European Social Fund (ESF), the European Regional Development Fund (ERDF), the European Agricultural Guidance and Guarantee Fund (EAGGF), the Integrated Mediterranean Programmes (IMPs), and the Research and Technological Development (RTD) Programmes.

Framework
Programme

The RTD programmes have developed out of early Community policies on Science and Technology. Important guidelines were adopted in 1974 which covered the coordination of national policies in science and technology, the participation of member states in the European Science Foundation, and the start of a Community programme of research in science and technology, including research on forecasting, assessment and methodology. A number of specialist committees were set up (with scientists as members) in order to advise on general and specific programmes. This resulted in the so called "Framework Programme" which consisted of a collection of coordinated RTD programmes. The first Framework Programme was drawn up for the period 1984-1987, the second for 1987-1991 and the third for 1990 to 1994 (overlapping the previous one). Most of the RTD programmes which are related to Biotechnology were set up under the second and third Framework Programmes. For example, the third Framework Programme was adopted by the Council in 1990 and had a duration of 5 years, a budget of 5.7 billion ECU and contained 15 specific RTD programmes in the following areas:

- information technologies;
- communications technologies;
- development of telematics systems;
- industrial and materials technologies;
- measurement and testing environment;
- marine science and technology;
- biotechnology;
- agricultural and agro-industrial research;
- life sciences and technologies for developing countries;
- non-nuclear energies;
- nuclear fission safety;
- controlled thermonuclear fusion;
- human capital and mobility.

The aim of the research programme in biotechnology is to reinforce basic biological knowledge as the common and integrated foundation for applications in agriculture, industry, health and the environment. The budget available to support the Biotechnology programme was 143 million ECU.

A list of programmes under which funding can be sought for Biotechnology research and training is given in Appendix 1.1

1.4.2 EC legislation on Health, Safety, Quality Control which regulate Biotechnology

EC legislation that affects biotechnology has arisen from a number of different areas within the Community. For example, laws on the environment first came into being during the First Programme (1973-1977) which was concerned with protecting the aquatic environment (by controlling pollutants), protecting the atmosphere (improving air quality standards), controlling noise pollution, controlling hazardous chemicals in the environment generally, and the management of disposal. Several directives were enacted during this period, but the most important guideline was that "the polluter pays for the prevention and elimination of environmental nuisance". However, one other main objective of this programme was the promotion of scientific research and the improvement of information, training and awareness of environmental problems.

the principle that "the polluter pays"

This is, therefore, a good example of where research on the one hand is coupled with guidelines or regulatory laws that are being proposed or enacted on the other.

The Second Programme (1977-1983) was mainly devoted to the implementation of the first programme. However, there was placed great emphasis on protection of the marine environment following the Amoco Cadiz oil-spill disaster.

The Third Programme (1983-1987) focussed on International co-operation in respect of protection of the aquatic environment. Community legislation to ensure that the movement of hazardous waste was supervised and controlled came into being as did new directives on the control of chemicals, dangerous substances and preparations.

The Fourth Programme (1988-1992) has continued to ensure that environmental laws have been central in the legislative process. It includes several new laws which affect disposal of waste, protection of wildlife, protection of forests and even directives to protect animals used for experimental purposes. During this time a number of advisory committees were established including a Committee on Waste Management, and a Scientific Advisory Committee on the Toxicity and Ecotoxicity of Chemical Compounds.

Consumer protection

During the course of the First, Second, Third and Fourth Programmes, parallel legislation was being proposed and implemented in the areas of Consumer protection. Amongst other things, consumer health and safety became an important subject, around which a number of important directives were made. For example, by the end of 1976 some 80 directives had been issued with regard to protecting consumer's health. These covered safety in a wide range of products including fruit, vegetables, foodstuffs and animal health and welfare, but also measures to protect against chemical solvents, paints, varnishes and household goods. In 1978, a Scientific Committee on Cosmetology was set up in order to provide the Commission with opinions on the toxicological effects of the use of cosmetics. This helped provide a sound scientific background to legislation. During the Second Programme, legislation centred around consumer protection in the field of services, whilst in the Third Programme, legislative action was initiated to deal with outstanding problems of co-ordination of consumer protection with other Community policies. The idea of developing Community standards of quality (eg. in the food sector or in the pharmaceutical sector) in order to give a degree of consumer protection was vigourously followed. In the Fourth Programme, this was continued and important directives were made including some which covered trade in hormone-treated animals, food labelling, cosmetics, foodstuffs, detergents and cleaning substances and aerosol dispensers. Examples of some particularly important health and safety directives and guidelines are given in later chapters of this source book. An outline of some of the more important legislation relating to biotechnology as a whole is given in chapter 3.

Appendix 1.1

1.1. Sources of European Funding for Biotechnology

1.1a BRIDGE (Biotechnology Research for Innovation, Development and Growth in Europe)

Scope: Research and training in a broad range of biotechnology techniques, information storage and dissemination, raising public awareness

Duration: 5 years, 1990-1994

EC contribution: 100 million ECU

Type of support: Training contracts, concerted actions

Note: This programme is now closed. The successor is BIOTECH.

1.1.b. BIOTECH

Scope: More orientated towards basic biology than the BRIDGE programme

Prenormative Research

Safety Assessment of New Techniques

Novel Products

Duration: 4 years 1992-1996

EC contribution: 164 million ECU

Type of support: Shared-cost research contracts, concerted actions, training contracts

1.1.c. FLAIR (Food Linked Agro-Industrial Research programme)

Scope: Assessment and enhancement of food quality and diversity

Food hygiene, saftey and toxicological aspects

Nutrition and wholesomeness aspects

Duration: 4 years 1989-1993

EC contribution: 25 million ECU

Type of support: Shared cost research contracts, concerted actions, scholarships

Note: Successor programme (Agro-Industrial Research) now in operation

1.1.d. Agriculture and Agro-Industrial Research

Scope: Agriculture, horticulture, forestry, aquaculture, food and non-food industries

Primary production, inputs processing of biological rawmaterials, end use products

Precompetitive research

Duration: 1991-1994

EC contribution: 309.67 million ECU

Type of support: Shared-cost research contracts, concerted actions

1.1.e. Non-Nuclear Energies

Scope: Models for energy and environment

Rational use of energy

Minimum emissions power production from fossil fuels

Renewable energies

Duration: 5 years 1990-1994

EC contribution: 157 million ECU

Type of support: Shared cost research contracts, concerted actions, scholarships

1.1.f. STD (Science and Technology for Development)

Scope: Tropical and subtropical agriculture

Medicine, health and nutrition in tropical and sub-tropical areas

Duration: 5 years 1990-1994

EC contribution: 111 million ECU

Type of support: Shared cost research contracts

1.1.g. MAST (Marine Science and Technology)

Scope: Basic and applied research in the marine sciences

Coastal zone science and engineering

Marine technology

Supporting initiatives

Large targeted projects

Duration: 3 years 1991-1994

EC contribution: 91.5 million ECU

Type of support: Shared cost research contracts, concerted actions, scholarships

1.1.h. EUREKA

Scope: EUREKA is a framework for industry-led projects aimed at producing high technology goods and services. EUREKA projects are nearer to the market than the EC funded research projects

Duration: on going since 1985

Type of support: self-funding or funding from private or public bodies

1.1.i. ERASMUS (European Action Scheme for the Mobility of University Students)

Scope: The ERASMUS programme promotes the mobility of students in the community and greater cooperation between universities. Students are given the opportunity to complete a recognized part of their study in another member state of the community

Duration: ongoing since 1985

EC contribution: 307.5 million ECU up to 1992

100 million ECU in 1992 -1993

Type of support: scholarships, inter-university co-operation programmes

1.1.j. Comett II (Community action programme in Education and Training for Technology)

Scope: Development and reinforcement of partnerships in the field of education between universities and business

Cross-border interchange of students and university graduates, scientists and specialists

Projects to promote continuing training in the technology sector and multimedia distance education

Complementary and associated initiatives

Duration: 1990-1994

EC contribution: 230 million ECU

Type of support: scholarships, courses, training programmes

1.1.k. EMBO (European Molecular Biology Organisation) research fellowships)

Object: Short-term fellowships to support 1-12 week visits to laboratories to carry out experiments with special techniques, collaborate, training etc. Long-term fellowships to support one year or two year fellowships to pursue collaborative research at foreign laboratories

Scope: Open to PhD status scientists in the field of molecular biology.

Tenable in laboratories in Israel and Europe

NOT tenable in industrial laboratories

Duration: Ongoing

Appendix 1.2

Addresses of the offices dealing with the Official Journal in EC - Member States

Belgium: Moniteur Belge, Rue de Louvain 40-42, B-1000 Brussels

Denmark: Statstidende, Otto Monsteds Gade 3, DK-1571, Kobenhavn V

Federal Republic of Germany: Bundesanzeiger Verlags-GmbH, Postfach 10 80 06, D-5000 KOLN 1

France: Journal Officiel de la Republique Francaise , 26 rue Desaix, F-75727 Paris

Greece: Ephimeris Kyvernisseos Ellinikis Dimokratias , (Official

Journal, Government Publications) Kapodistriou 34, Athens

Ireland: Iris Oificuil, Stationery Office, Biship Street, Dublin 8

Italy: Gazetta Ufficiale della repubblica Italiana, Istituto Polografico e Zecca dello Stato Piazza G. Verdi 10, I-001j98 Roma

Luxembourge: Memorial Service Central de Legislation, Rue du Marche aux Herbes, Luxembourg

Netherlands: Nederlandse Staatscourant, Postbus 20014, NL-2500 EA DEN HAAG

United Kingdom: The London Gazette, HMSO Publication Centre, 51 Nine Elms Lane, London SW8 2DR

Directives, Decisions and Regulations (and how to find them)

Directives, Decisions and Regulations (and how to find them)

2.1 Introduction

In this chapter we look at the structure of Directives, Decisions and Regulations, taking real examples from the Official Journal and describing the general parts that make up these documents. We begin with a Directive. The example shown is of direct relevance to biotechnology since it deals with protection of workers from exposure to biohazards; for this reason it is shown in its entirety. The structure of the Directive will be compared to that of a Decision and that of a Regulation again choosing real examples from the OJ.

*amendments
supplements
and
replacements
are made to
existing laws*

The law is, however, never static. New proposals are being shaped into final legislation, new Directives are made to extend the scope of previous Directives; amendments, supplements and replacements are made to existing laws. Sometimes laws are challenged through the European Court which may influence the enactment of existing legislation. This chapter will help to describe the nature of amendments, supplements and replacements and suggests which sources of information are available for students to find out for themselves which EC regulations, amendments and supplements are currently in force and which relate to biotechnology.

This chapter does not describe actual national legislation derived from or encompassing EC Directives relating to biotechnology. For this information you will have to turn to later chapters.

2.2 Structure of a Directive

Table 1.4 of chapter 1 describes the difference between Regulations, Directives, Decisions, Recommendations and Opinions. All are published in the Official Journal of the European Communities. If we take the Directive, published in 1990 "on the protection of workers from risks related to biological agents at work" as an example (see Figure 2.1), the general structure of a Directive can be explained. Do not attempt to read right the way through Figure 2.1 at this stage. As you read through this text, we will draw your attention to particular points about the structure of this Directive.

*OJ reference
number*

The first thing that we note in our example (printed across the top) is the date of publication and the journal title. The number in the topright hand corner is composed of a letter (L in this particular case), a number (the OJ document number) and the page number. This collectively comprises the OJ reference number for this document and should be referred to as the publication reference. The 'L' series denotes that it has to do with legislation. There is also a 'C' series which is concerned with information and notices (in particular legal notices from the Court of Justice).

In the middle of the document, towards the top is a roman numeral. If this is a 'I' it means that the legislation is concerned with acts whose publication is obligatory. In our example the number is 'II' which indicates legislation whose publication is not obligatory. Below this appears the word "Council" and this indicates which EC body has issued the act in question. (This could have been Commission, European Parliament, Court of Justice, Court of Auditors, Economic and Social Committee etc., depending on the nature of the acts or information being published). Referring back to our example, it then states more directly that it is a "Council Directive" and gives the date on which the Directive was issued. The title of the legislation is then given. In this particular piece of legislation the title is complex since it refers to previous legislation. This can be ignored for the moment.

The next point to note is the number ("90/679/EEC" in our particular example) which is the EEC's reference number for this document and it must not be confused with the OJ publication number (which always starts with an 'L' or a 'C'). The first number (90) relates to the year that the piece of legislation was enacted (1990 in our case). The second number (697) is the serial number. It should be noted that for regulations, the serial number comes first with the year following.

Finally comes the actual text of the Directive. These always start with introductory provisions and make reference to the legal basis of the act in question (ie having regard to the Treaty establishing the European Economic Community). Further introductory paragraphs continue to indicate the legal basis of the act, its relationship to other acts, its legal development, its scope and its applicability (ie. its general philosophy). Each paragraph starts with the word "Whereas" and sets out the principles, or indicates the pertinence of the legislation (ie. why it is needed).

text divided into articles and sections or chapters

Following the introduction, comes the main body of the legislation. This is broken down into a series of Articles which are numbered progressively. In some cases (as in our example) the body is also broken down into "Sections" or "Chapters". However, each piece of legislation is different with respect to whether it has sections or not and as to how many articles it contains.

Annexes to Directives

Following the main body of the Directive there is a final part headed "Annex". In our example there are 6 annexes. It is the purpose of the Annex to give further information, such as lists (of products, of chemicals, of methods, of measures), or other complementary information such as exceptions, inclusions, indications, limits, etc.

It should be noted that where the legislation makes reference to previous acts it refers to these within the body of the text by their title, article number of the act, and/or EEC document number. It also indicates where these are published in the Official Journal by showing a small superscript number (index number) which refers to the corresponding number at the foot of the page. This now shows the OJ publication reference number where the particular act in question can be found. (Sometimes a non-OJ bibliographic reference may be found if it is pertinent).

We can now explain the full title which is "on the protection of workers from risks related to exposure to biological agents at work (seventh individual Directive within the meaning of Article 16 (1) of Directive 89/391/EEC)". The reference is to a previous Directive (89/391/EEC) published on June 12, 1989 which is titled "on the introduction of measures to encourage improvements in the safety and health of workers at work". In particular, Article 16 (1) of the earlier Directive relates to the later Directive.

Figure 2.1.
Example of an EC Directive

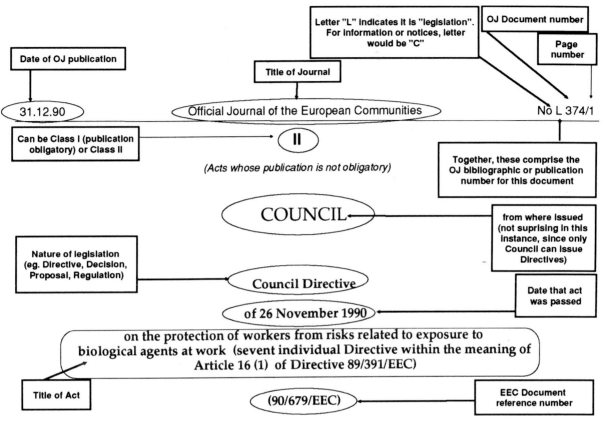

Date of OJ publication

Letter "L" indicates it is "legislation". For information or notices, letter would be "C"

OJ Document number

Page number

Title of Journal

31.12.90   Official Journal of the European Communities   No L 374/1

Can be Class I (publication obligatory) or Class II

II

Together, these comprise the OJ bibliographic or publication number for this document

(Acts whose publication is not obligatory)

COUNCIL

from where issued (not suprising in this instance, since only Council can issue Directives)

Nature of legislation (eg. Directive, Decision, Proposal, Regulation)

Council Directive

Date that act was passed

of 26 November 1990

on the protection of workers from risks related to exposure to biological agents at work (sevent individual Directive within the meaning of Article 16 (1) of Directive 89/391/EEC)

Title of Act

(90/679/EEC)

EEC Document reference number

THE COUNCIL OF THE EUROPEAN COMMUNITIES,

Having regard to the Treaty establishing the European Economic Community, and in particular Article 118a thereof,

Having regard to the proposal from the Commission (^{1}), drawn up after consulting the Advisory Committee on Safety, Hygiene and Health Protection at Work,

In cooperation with the European Parliament (^{2})

Having regard to the Opinion of the Economic and Social Committee(^{3}),

Whereas Article 118a of the Treaty provides that the Council shall adopt, by means of Directives, minimum requirements in order to encourage improvements, especially in the working environment, so as to guarantee better protection of the health and safety of workers;

Whereas that Article provides that such Directives shall avoid imposing administrative, financial and legal constraints in a way which would hold back the creation and development of small and medium-sized undertakings;

Wheras the Council Resolution of 27 February 1984 on a second action programme of the European Communities on safety and health at work (^{4}) provides for the development of protective measures for workers exposed to dangerous agents;

Whereas the communication from the Commission on its programme concerning safety, hygiene and health at work (^{5}) provides for the adoption of Directives to guarantee the safety and health of workers;

Whereas compliance with the minimum requirements designed to guarantee a better standard of safety and health as regards the protection of workers from the risks related to exposure to biological agents at work is essential to ensure the safety and health of workers;

Whereas this Directive is an individual Directive within the meaning of Article 16 (1) f Council Directive 89/391/EEC of 12 June 1989 on the introduction of measures to encourage improvements in the safety and health of workers

Introduction and introductory provisions

(^{3}) OJ No C 28, 3. 2 1988, p. 1.
(^{6}) OJ No L 183, 29, 6. 1989, p. 1.

(^{1}) OJ No C 150, 8. 6. 1988, p. 6.
(^{2}) OJ No C 158, 26. 6. 1989, p. 92.
(^{3}) OJ No C 56, 6. 3. 1989, p. 38.
(^{4}) OJ No C 67, 8. 3. 1984, p. 2.

References to previous acts mentioned in the text. All are OJ references in this particular example at work (6); whereas the provisions of that Directive are therefore fully applicable to the exposure of workers at work (6); whereas the provisions of that Directive are therefore fully applicable to the exposure of workers to biological agents, without prejudice to more stringent and/or specific provisions contained in the present Directive;

Whereas more precise knowledge of the risks involved in exposure to biological agents at work can be obtained through the keeping of records;

Whereas employers must keep abreast of new developments in technology with a view to improving the protection of workers' health and safety;

Whereas preventive measures should be taken for the protection of the health and safety of workers exposed to biological agents;

Whereas this Directive constitutes a practical aspect of the realization of the social dimension of the internal market;

Whereas, pursuant to Decision 74/325/EEC ([1]), the Advisory Committee on Safety, Hygiene and Health Protection at Work is consulted by the Commission on the drafting of proposals in this field,

HAS ADOPTED THIS DIRECTIVE:

> **MAIN "BODY" OF ACT STARTS FROM HERE: CONTINUES UNTIL START OF ANNEXES**

SECTION I
GENERAL PROVISIONS

Article 1
Objective

1. This Directive, which is the seventh individual Directive within the meaning of Article 16 (1) of Directive 89/391/EEC, has as its aim the protection of workers against risks to their health and safety, including the prevention of such risks, arising or likely to arise from exposure to biological agents at work.

It lays down particular minimum provisions in this area.

2. Directive 89/391/EEC shall apply fully to the whole area referred to in paragraph 1, without prejudice to more stringent and/or specific provisions contained in this Directive.

3. This Directive shall apply without prejudice to the provisions of Council Directive 90/219/EEC of 23 April 1990 on the contained use of genetically modified micro-organisms ([2]) and of Council Directive 90/220/EEC of 23 April 1990 on the deliberate release into the environment of genetically modified organisms ([3]).

Article 2
Definitions

For the purpose of this Directive:

(a) 'biological agents' shall mean micro-organisms, including those which have been genetically modified, cell cultures and human endoparasites, which may be able to provoke any infection, allergy or toxicity;

(1) OJ No L 185, 9. 7. 1974, p. 15.
(2) OJ No L 117, 8. 5. 1990, p. 11
(3) OJ No L 117, 8. 5. 1990, p. 15.

(b) 'micro-organism' shall mean a microbiological entity, cellular or non-cellular, capable of replication or of transferring genetic material;

(c) 'cell culture' shall mean the in-vitro growth of cells derived from multicellular organisms;

(d) 'biological agents' shall be classified into four risk groups, according to their level of risk of infection:

1. group 1 biological agent means one that is unlikely to cause human disease;

2. group 2 biological agent means one that can cause human disease and might be a hazard to workers; it is unlikely to spread to the community; there is usually effective prophylaxis or treatment available;

3. group 3 biological agent means one that can cause severe human disease and present a serious hazard to workers; it may present a risk of spreading to the community, but there is usually effective prophylaxis or treatment available;

4. group 4 biological agent means one that causes severe human disease and is a serious hazard to workers; it may present an high risk of spreading to the community; there is usually no effective prophylaxis or treatment available.

Article 3
Scope - Determination and assessment of risks

1. This Directive shall apply to activities in which workers are or are potentially exposed to biological agents as a result of their work.

2. (a) In the case of any activity likely to involve a risk of exposure to biological agents, the nature, degree and duration of workers' health or safety and to lay down the measures to be taken.

(b) In the case of activities involving exposure to several groups of biological agents,, the risk shall be assessed on the basis of the danger presented by all hazardous biological agents present.

(c) The assessment must be renewed regularly and in any event when any change occurs in the conditions which may affect workers' exposure to biological agents.

(d) The employer must supply the competent authorities, at their request, with the information used for making the assessment.

3. The assessment referred to in paragraph 2 shall be conducted on the basis of all available information including:

- classification of biological agents which are or may be a hazard to human health, as referred to in Article 18;

- recommendations from a competent authority which indicate that the biological agent should be controlled in order to protect workers' health when workers are or may be exposed to such a biological agent as a result of their work;

- information on diseases which may be contracted as a result of the work of the workers;

- potential allergenic or toxigenic effects as a result of the work of the workers;

- knowledge of a disease from which a worker is found to be suffering and which has a direct connection with his work.

Article 4

Application of the various Articles in relation to assessment of risks

1. If the results of the assessment referred to in Article 3 show that the exposure and/or potential exposure is to a group 1 biological agent, with no identifiable health risk to workers, Articles 5 to 17 and Article 19 shall not apply.

However, point 1 of Annex VI should be observed.

2. If the results of the assessment referred to in Article 3 show that the activity does not involve a deliberate intention to work with or use a biological agent but may result in the workers being exposed to a biological agent, as in the course of the activities for which an indicative list is given in Annex I. Articles 5, 7, 8, 10, 11, 12, 13 and 14 shall apply unless the results of the assessment referred to in Article 3 show them to be unnecessary.

NOTE: ARTICLE NUMBERS CONTINUE THROUGHOUT DESPITE NEW SECTION

SECTION II
EMPLOYERS' OBLIGATIONS

Article 5

Replacement

The employer shall avoid the use of a harmful biological agent if the nature of the activity so permits, by replacing it with a biological agent which, under its conditions of use, is not dangerous or is less dangerous to workers' health, as the case may be, in the present state of knowledge.

Article 6

Reduction of risks

1. Where the results of the assessment referred to in Article 3 reveal a risk to workers' health or safety, workers; exposure must be prevented.

2. Where this is not technically practicable, having regard to the activity and the risk assessment referred to in Article 3, the risk of exposure must be reduced to as low a level as necessary in order to protect adequately the health and safety of the workers concerned, in particular by the following measures which are to be applied in the light of the results of the assessment referred to in Article 3:

(a) keeping as low as possible the number of workers exposed or likely to be exposed;

(b) design of work processes and engineering control measures so as to avoid or minimize the release of biological agents into the place of work;

(c) collective protection measures and/or, where exposure cannot be avoided by other means, individual protection measures;

(d) hygiene measures compatible with the aim of the prevention or reduction of the accidental transfer or release of a biological agent from the workplace:

(e) use of the biohazard sign depicted in Annex II and other relevant warning signs;

(f) drawing up plans to deal with accidents involving biological agents;

(g) testing, where it is necessary and technically possible, for the presence, outside the primary physical confinement, of biological agents used at work;

(h) means for safe collection, storage and disposal of waste by workers, including the use of secure and identifiable containers, after suitable treatment where appropriate;

(i) arrangements for the safe handling and transport of biological agents within the workplace.

Article 7

Information for the competent authority.

Where the results of the assessment referred to in Article 3 reveal a risk to workers' health or safety, employers shall, when requested, make available to the competent authority appropriate information on:

- the results of the assessment;

- the activities in which workers have been exposed or may have been exposed to biological agents;

- the number of workers exposed;

- the name and capabilities of the person responsible for safety and health at work;

- the protective and preventive measures taken, including working procedures and methods;

 an emergency plan for the protection of workers from exposure to a group 3 or a group 4 biological agent which might result from a loss of physical containment.

2. Employers shall inform forthwith the competent authority of any accident or incident which may have resulted in the release of a biological agent and which could cause severe human infection and/or illness.

3. The list referred to in Article 11 and the medical record referred to in Article 14 shall be made available to the competent authority in cases where the undertaking ceases activity, in accordance with national laws and/or practice.

Article 8

Hygiene and individual protection

1. Employers shall be obliged, in the case of all activities for which there is a risk to the health or safety of workers due to work with biological agents, to take appropriate measures to ensure that:

(a) workers do not eat or drink in working areas where there is a risk of contamination by biological agents;

(b) workers are provided with appropriate protective clothing or other appropriate special clothing;

(c) workers are provided with appropriate and adequate washing and toilet facilities which may include eye washes and/or skin antiseptics;

(d) any necessary protective equipment is:

- properly stored in a well-defined place;

 checked and cleaned if possible before, and in any case after, each use;

- is repaired, where defective, or is replaced before further use;

(e) procedures are specified for taking, handling and processing samples of human or animal origin.

2. (a) Working clothes and protective equipment, including protective clothing referred to in paragraph 1, which may be contaminated by biological agents, must be removed on

leaving the working area and, before taking the measures referred to in subparagraph (b), kept separately from other clothing.

(b) The employer must ensure that such clothing and protective equipment is decontaminated and cleaned or, if necessary, destroyed.

3. Workers may not be charged for the cost of the measures referred to in paragraphs 1 and 2.

Article 9

Information and training of workers

1. Appropriate measures shall be taken by the employer to ensure that workers and/or any workers' representatives in the undertaking or establishment receive sufficient and appropriate training, on the basis of all available information, in particular in the form of information and instructions, concerning:

(a) potential risks to health;

(b) precautions to be taken to prevent exposure:

(c) hygiene requirements;

(d) wearing and use of protective equipment and clothing;

(e) steps to be taken by workers in the case of incidents and to prevent incidents.

2. The training shall be:

- given at the beginning of work involving contact with biological agents,

- adapted to take account of new or changed risks, and

- repeated periodically if necessary.

Article 10

Worker information in particular cases

1. Employers shall provide written instructions at the workplace and, if appropriate, display notices which shall, as a minimum, include the procedure to be followed in the case of:

- a serious accident or incident involving the handling of a biological agent;

- handling a group 4 biological agent.

2. Workers shall immediately report any accident or incident involving the handling of a biological agent to the person in charge or to the person responsible for safety and health at work.

3. Employers shall immediately report any accident or incident involving the handling of a biological agent to the person in charge or to the person responsible for safety and health at work.

3. Employers shall inform forthwith the workers and/or any workers representatives of any accident or incident which may have resulted in the release of a biological agent and which could cause severe human infection and/or illness.

In addition, employers shall inform the workers and/ or any workers' representatives in the undertaking or establishment as quickly as possible when a serious accident or incident occurs, of the causes thereof, and of the measures taken or to be taken to rectify the situation.

4. Each worker shall have access to the information on the list referred to in Article 11 which relates to him personally.

5. Workers and/or any workers' representatives in the undertaking or establishment shall have access to anonymous collective information.

6. Employers shall provide workers and/or their representatives, at their request, with the information provided for in Article 7 (1).

Article 11

List of exposed workers

1. Employers shall keep a list of workers exposed to group 3 and, whenever possible, the biological agent to which they have been exposed, as well as records of exposures, accidents and incidents, as appropriate.

2. The list referred to in paragraph 1 shall be kept for at least 10 years following the end of exposure, in accordance with national laws and/or practice.

In the case of those exposures which may result in infections:

- with biological agents known to be capable of establishing persistent or latent infections,

- that, in the light of present knowledge, are undiagnosable until illness develops many years later,

- that have particularly long incubation periods before illness develops,

- that result in illnesses which recrudesce at times over a long period despite treatment, or

- that may have serious long-term sequelae,

the list shall be kept for an appropriately longer time up to 40 years following the last known exposure.

3. The doctor referred to in Article 14 and/or the competent authority for health and safety at work, and any other person responsible for health and safety at work, shall have access to the list referred to in paragraph 1.

Article 12

Consultation and participation of workers

Consultation and participation of workers and/or their representatives in connection with matters covered by this Directive including the Annexes shall take place in accordance with Article 11 of Directive 89/391/EEC.

Article 13

Notification to the competent authority

1. Prior notification shall be made to the competent authority of the use for the first time of:

- group 2 biological agents.

_ group 3 biological agents,

- group 4 biological agents.

The notification shall be made at least 30 days before the commencement of the work.

Subject to paragraph 2, prior notification shall also be made of the use for the first time of each subsequent group 4 biological agent and of any subsequent new group 3 biological agent where the employer himself provisionally classifies that biological agent.

2. Laboratories providing a diagnostic service in relation to group 4 biological agents shall be required only to make an initial notification of their intention.

3. Renotification must take place in any case where there are substantial changes of importance to safety or health at work to processes and/or procedures which render the notification out of date.

4. The notification referred to in this Article shall include:

(a) the name and address of the undertaking and/or establishment;

(b) the name and capabilities of the person responsible for safety and health at work;

(c) the results of the assessment referred to in Article 3;

(d) the species of the biological agent;

(e) the protection and preventive measures that are envisaged.

SECTION III
MISCELLANEOUS PROVISIONS
Article 14
Health surveillance

1. The Member States shall establish, in accordance with national laws and practice, arrangements for carrying out relevant health surveillance of workers for whom the results of the assessment referred to in Article 3 reveal a risk to health or safety.

2. The arrangements referred to in paragraph I shall be such that each worker shall be able to undergo, if appropriate, relevant health surveillance:

- prior to exposure,

- at regular intervals thereafter.

Those arrangements shall be such that it is directly possible to implement individual and occupational hygiene measures.

3. The assessment referred to in Article 3 should identify those workers for whom special protective measures may be required.

When necessary, effective vaccines should be made available for those workers who are not already immune to the biological agent to which they are exposed or are likely to be exposed.

If a worker is found to be suffering from an infection and/or illness which is suspected to be the result of exposure, the doctor or authority responsible for health surveillance of workers shall offer such surveillance to other workers who have been similarly exposed.

In that event, a reassessment of the risk of exposure shall be carried out in accordance with Article 4.

4. In cases where health surveillance is carried out, an individual medical record shall be kept for at least 10 years following the end of exposure, in accordance with national laws and practice.

In the special cases referred to in Article 11(2) second subparagraph, an individual medical record shall be kept for an appropriately longer time up to 40 years following the last known exposure.

5. The doctor or authority responsible for health surveillance shall propose any protective or preventive measures to be taken in respect of any individual worker.

6. Information and advice must be given to workers regarding any health surveillance which they may undergo following the end of exposure.

7. In accordance with national laws and/or practice:

- workers shall have access to the results o the health surveillance which concern them, and

- the workers concerned or the employer may request a review of the results of the health surveillance8. Practical recommendations for the health surveillance of workers are given in Annex IV.

9. All cases of diseases or death identified in accordance with national laws and/or practice as resulting from occupational exposure to biological agents shall be notified to the competent authority.

Article 15

Health and veterinary care facilities other than diagnostic laboratories

1. For the purpose of the assessment referred to in Article 3, particular attention should be paid to:

(a) uncertainties about the presence of biological agents in human patients or animals and the materials and specimens taken from them;

(b) the hazard represented by biological agents known or suspected to be present in human patients or animals and materials and specimens taken from them;

(c) the risks posed by the nature of the work.

2. Appropriate measures shall be taken in health and veterinary care facilities in order to protect the health and safety of the workers concerned.

The measures to be taken shall include in particular:

(a) specifying appropriate decontamination and disinfection procedures, and

(b) implementing procedures enabling contaminated waste to be handled and disposed of without risk.

3. In isolation facilities where there are human patients or animals who are, or who are suspected of being, infected with group 3 or group 4 biological agents, containment measures shall be selected from those in Annex V column A, in order to minimize the risk of infection.

Article 16

Special measures for industrial processes, laboratories and animal rooms

1. The following measures must be taken in laboratories, including diagnostic laboratories, and in rooms for laboratory animals which have been deliberately infected with group 2, 3 or 4 biological agents or which are or are suspected to be carriers of such agents:

(a) Laboratories carrying out work which involves the handling of group 2, 3 or 4 biological agents for research, development, teaching or diagnostic purposes shall determine the containment measures in accordance with Annex V, in order to minimize the risk of infection.

(b) Following the assessment referred to in Article 3, measures shall be determined in accordance with Annex V, after fixing the physical containment level required for the biological agents according to the degree of risk.

Activities involving the handling of a biological agent must be carried out:

- only in working areas corresponding to at least containment level 2, for a group 2 biological agent;

- only in working areas corresponding to at least containment level 3, for a group 3 biological agent;

- only in working areas corresponding to at least containment level 4, for a group 4 biological agent.

(c) Laboratories handling materials in respect of which there exist uncertainties about the presence of biological agents which may cause human disease but which do not have as their aim working with biological agents as such (i.e. cultivating or concentrating them) should adopt containment level 2 at least. Containment levels 3 or 4 must be used, when appropriate, where it is known or it is suspected that they are necessary, except where guidelines provided by the competent national authorities show that, in certain cases, a lower containment level is appropriate.

2. The following measures concerning industrial processes using group 2, 3 or 4 biological agents must be taken:

(a) The containment principles set out in the second subparagraph of paragraph 1 (b) should also apply to industrial processes on the basis of the practical measures and appropriate procedures given in Annex VI.

(b) In accordance with the assessment of the risk linked to the use of group 2, 3 or 4 biological agents, the competent authorities may decide on appropriate measures which must be applied to the industrial use of such biological agents.

(c) For all activities covered by this Article where it has not been possible to carry out a conclusive assessment of a biological agent but concerning which it appears that the use envisaged might involve a serious health risk for workers, activities may only be carried out in workplace where the containment level corresponds at least to level 3.

Article 17

Use of data

The Commission shall have access to the use made by the competent national authorities of the information referred to in Article 14 (9).

Article 18

Classification of biological agents

1. In accordance with the procedure laid down in Article 118a of the Treaty, the Council shall adopt within six months of the date of implementation given in Article 20(1) a first list of group 2, group 3 and group 4 biological agents for Annex III

2. Community classification shall be on the basis of the definitions in Article (d) points 2 to 4 (groups 2 to 4).

3. Pending Community classification Member States shall classify biological agents that are or may be a hazard to human health on the basis of the definition in Article 2 (d), it must be classified in the highest risk group among the alternatives.

Article 19

Annexes

Purely technical adjustments to the Annexes in the light of technical progress, changes in international regulations or specifications and new findings in the field of biological agents shall be adopted in accordance with the procedure laid down in Article 17 of Directive 89/391/EEC

Article 20

Final provisions

1. Member States shall bring into force the laws, regulations and administrative provisions necessary to comply with this Directive not later than three years after the notification of this Directive ([1]). They shall forthwith inform the Commission thereof.

However, in the case of the Portuguese Republic, the time limit referred to in the first subparagraph shall be five years.

> **Time-limit by which Member States have to implement the Directive and after which the Directive or some of its provisions will have direct effect**

2. Member States shall communicate to the Commission the provisions of national law already adopted or which they adopt in the field governed by this Directive.

Article 21

This Directive is addressed to the Member States.

Done at Brussels. 26 November 1990

For the Council

The President

C.; DONAT CATTIN

> **Name of the Minister concerned of the government which at that time was presiding over the Council of Ministers (in this example, Italy**

ANNEX I

INDICATIVE LIST OF ACTIVITIES

(Article (2))

1. Work in food production plants.

2. Work in agriculture.

3. Work activities where there is contact with animals and/or products of animal orgin.

4. Work in health care, including isolation and post mortem units.

5. Work in clinical, veterinary and diagnostic laboratories, excluding diagnostic microbiological laboratories.

6. Work in refuse disposal plants.

7. Work in sewage purification installations.

ANNEX II

BIOHAZARD SIGN

(Article 6 (2) (e)

ANNEX III

Community Classification (Articles 18 and 2 (d)

for the record

ANNEX IV

PRACTICAL RECOMMENDATIONS FOR THE HEALTH SURVEILLANCE OF WORKERS

(article 14 (8))

1. The doctor and/or the authority responsible for the health surveillance of workers exposed to biological agents must be familiar with the exposure conditions or circumstances of each worker.

2. Health surveillance of workers must be carried out in accordance with the principles and practices of occupational medicine: it must include at least the following measures:

- keeping records of a worker's medical and occupational history;

- a personalized assessment of the workers' state of health;

- where appropriate, biological monitoring, as well as detection of early and reversible effects.

Further tests may be decided upon for each worker when he is the subject of health surveillance, in the light of the most recent knowledge available to occupational medicine.

ANNEX V

INDICATIONS CONCERNING CONTAIMENT MEASURES AND CONTAINMENT LEVELS

(Articles 15 (3) and 16 (1) (a) and (b))

Preliminary note

The measures contained in this Annex shall be applied according to the nature of the activities the assessment of risk to workers, and the nature of the biological agent concerned.

A. Containment measures	B. Containment levels		
	2	3	4
1. The workplace is to be separated from any other activities in the same building	No	Recommended	Yes
2. Input air and extract air to the workplace are to be filteredusing (HEPA) or likewise	No	Yes, on extract air	Yes, on input and extract air
3. Access is to be restricted to nominated workers only	Recommended	Yes	Yes, via airlock
4. The workplace is to be sealable to permit disinfection	No	Recommended	Yes
5. Specified disinfection procedures	Yes	Yes	Yes
6. The workplace is to be maintained at an air pressure negative to atmosphere	No	Recommended	Yes
7. Efficient vector control e.g. rodents and insects	Recommended	Yes	Yes

ANNEX VI

CONTAINMENT FOR INDUSTRIAL PROCESSES

(Article 4 (1) and Article 16 (2) (a))

Group 1 biological agents

For work with group 1 biological agents including life attenuated vaccines, the principles of good occupational safety and hygiene should be observed.

Group 2,3 and 4 biological agents

It may be appropriate to select and combine containment requirements from different categories below on the basis of a risk assessment related to any particular process or part of a process.

A. Containment measures	A Containment levels		
	2	3	4
1. Viable organisms should be handled in a system which physically separates the process from the environment	Yes	Yes	Yes
2. Exhaust gases from the closed system should be treated so as to:	Minimize release	Prevent release	Prevent release
3. Sample collection, addition of materials to a closed system and transfer of viable organisms to another closed system, should be performed so as to:	Minimize release	Prevent release	Prevent release

2.3 Structure of a Decision

The example shown (Figure 2.2) has a similar structure to the previous example given for a Directive. However, in this case the legislation is issued by the Commission, and because it is a Decision, it is binding towards those that it is addressed to. In our example it is addressed to the UK as a whole. In practice this means all those involved in animal embryo work in cattle and the various UK authorities responsible for regulating this activity and who have a responsibility for export licences, health certificates etc for animals and animal embryos. The example given is short (but not all Decisions are short); the main body has no sections, just 5 Articles. Moreover, there are no Annexes on this particular piece of legislation.

2.4 Structure of a Regulation

This follows the same general structure of a Directive or a Decision. However, Regulations are binding on all member states and therefore tend to be concerned with the unification of the member-states' legislations (where this is shown to be necessary). More often than not regulations relate directly or indirectly to fair trade between EC Member States so that the internal market of the EC can be progressively established. The truncated example given (Figure 2.3) sets out the Community methods to be adopted for the measurement and analysis of alcohol in wine. This allows for comparisons to be made across the wine sector of the Member States.

Figure 2.2
Example of an EC Decision

II

(Acts whose publication is not obligatory

COMMISSION

COMMISSION DECISION

of 14th May 1992

concerning certain protection measures relating to bovine embryos in respect of bovine spongiform encephalopathy (BSE) in the United Kingdom

(92/290/EEC)

THE COMMISSION OF THE EUROPEAN COMMUNITIES,

Having regard to the Treaty establishing the European Economic Community,

Having regard to Council Directive 90/425/EEC of 26 June 1990 concerning veterinary and zoo technical checks applicable in intra-Community trade in certain live animals and products with a view to the completion of the internal market (^{1}), as last amended by Directive 91/628/EEC (^{2}), and in particular Article 10 thereof,

Whereas several outbreaks of bovine spongiform encephalopathy have occurred throughout the territory of the United Kingdom;

Whereas this disease can be considered to be a serious contagious or infectious animal disease whose presence may constitute a danger to cattle in other Member States;

Whereas previously it was considered that trade in bovine embryos presented only a negligible risk of endangering the health of livestock in Member States with respect to bovine spongiform encephalopathy; whereas, however, recent evidence relating to sheep experimentally infected with scrapie now suggests that bovine embryos may represent more than a negligible risk;

Whereas, however, the risk is considered to exist only for embryos from donors born before 18 July 1988 or born to infected cows, in view of the epidemiology and pathogenesis of the disease;

Whereas it is necessary to apply certain measures to intra-Community trade in bovine embryos from the United Kingdom to overcome this risk;

Whereas the authorities of the United Kingdom have undertaken to implement national measures necessary to guarantee the efficient implementation of this Decision;

Whereas it is also necessary to ensure that no embryo derived from any suspected or confirmed case is entered into trade;

Whereas it is also necessary to ensure that no embryo derived from any suspected or confirmed case is entered into trade;

Whereas Council Directive 89/556/EEC of 25 September 1989 on animal health conditions governing intra-Community trade in and importation from third countries of embryos of

domestic animals of the bovine species (^{3}), as amended by Directive 90/425/EEC, makes provisions for a health certificate to accompany bovine embryos in intra-Community trade; whereas the certificate must be modified in respect of bovine embryos from the United Kingdom;

Whereas the measures provided for in this Decision are in accordance with the opinion of the Standing Veterinary Committee,

HAS ADOPTED THIS DECISION:

Article 1

Member States shall not send to other Member States embryos of the domestic bovine species derived from females in which, at the time of sending, bovine spongiform encephalopathy is suspected or confirmed.

Article 2

1. The United Kingdom shall not send to other Member states embryos of the domestic bovine species derived from females which:

were born before 18 July 1988,

are the offspring of females in which bovine spongiform encephalopathy is suspected or confirmed.

2. The provisions of paragraph 1 shall not apply to embryos derived from females born outside the United Kingdom and subsequently introduced into the United Kingdom after 18 July 1988.

3. The United Kingdom shall make full use of records to guarantee identification of donors and embryos.

Article 3

The health certificate provided for in Annex C to Directive 89/556/EEC accompanying embryos sent from the United Kingdom shall be completed by the following:

'Embryos in accordance with Commission Decision 92/290/EEC concerning bovine spongiform encephalopathy'

Article 4

Member States shall amend the measures which they apply to trade so that they comply with this Decision 15 days after its notification. They shall immediately inform the Commission thereof.

(^{1}) OJ No L 224, 18. 8. 1990, p. 29.
(^{2}) OJ No L 340, 11. 12. 1991, p. 17.

(^{3}) OJ No L 302, 19. 10. 1989, p. 1l

Article 5

This Decision is addressed to the Member States.

Done at Brussels, 14 May 1992.

For the Commission

Ray Mac Sharry

Member of the Commission

Figure 2.3
Example of an EC
Regulation(truncated version,
does not include the annex)

COMMISSION REGULATION (EEC) No 1238/92

of 8 May 1992

determining the Community methods applicable in the wine sector for the analysis of neutral alcohol

THE COMMISSION OF THE EUROPEAN COMMUNITIES,

Having regard to the Treaty establishing the European Economic Community,

Having regard to Council Regulation (EEC) No 822/87 of 16 March 1987 on the common organization of the market in wine (^{1}), as last amended by Regulation (EEC) No 1734/91 (^{2}), and in particular Articles 35 (8), 36 (6), 38 (5), 39 (9), 41 (10) and 42 (6),

Whereas, under Council Regulation (EEC) No 2046/89 of 19 June 1989 laying down general rules for distillation operations involving wine and the by products of wine making (^{3}), neutral alcohol obtained by distillation operations in the wine-growing sector must be as defined in the Annex to that Regulation on the basis of criteria relating to its composition; whereas Community methods of analysis should be adopted in order to check whether the criteria have been complied with;

Whereas these methods must be binding for all commercial transactions and control operation; whereas, in view of the restricted opportunities for trade, a limited number of general methods permitting a rapid and sufficiently accurate analysis of the required components of the neutral alcohol should be adopted;

whereas the Community methods of analysis adopted should be generally recognized in order to ensure that they are supplied in a uniform basis;

Whereas the current Community methods of analysis for neutral alcohol in the wine sector were adopted by Commission Regulation (EEC) No 3590/83 (^{4}); whereas scientific progress necessitates the replacement of some methods with more suitable ones, the amendment of other methods and the introduction of new ones; whereas in view of the large number and complexity of these changes, all the methods of analysis should be incorporated into a new Regulation and Regulation (EEC) No 3590/83 should be replaced;

Whereas the terms used for the repeatability and comparability of the results obtained with these methods should be defined so that the results obtained in application of the methods of analysis listed in Article 74 of Regulation (EEC) No 822/87 can be compared;

Whereas the measures provided for in this Regulation are in accordance with the opinion of the Management Committee for Wine,

HAS ADOPTED THIS REGULATION:

Article 1

1. The Community methods for the analysis of neutral alcohol as defined in the Annex to Regulation (EEC) No 2046/89 shall be as set out in the Annex to this Regulation.

2. The methods of analysis specified in paragraph 1 shall apply to neutral alcohol obtained by the distillation operations provided for in Regulation (EEC) No 822/87.

Article 2

For the purposes of applying this Regulation:

(a) the repeatability shall be the value below which the absolute difference between the two single tests results obtained using tests conducted under the same conditions (same operator, same apparatus, same laboratory and a short interval of time) may be expected to lie within a specified probability;

(b) the comparability shall be the value below which the absolute difference between twos single test results obtained under different conditions (different operators different apparatus and/or different laboratories and or different time) may be expected to lie within a specified probability.

The term single test result shall be the value obtained when the standardized test method is applied fully and once to a single sample. Unless otherwise stated, the probability shall be 95%.

Article 3

Regulation (EEC) No 3590/83 is hereby repealed.

Article 4

This Regulation shall enter into force on the third day after its publication in the Official Journal of the European Communities.

This Regulation shall be binding in its entirety and directly applicable in all Member States.

Done at Brussels, 8 May 1992.

For the Commission

Ray MAC SHARRY

Member of the Commission

(^{3}) OJ No L 302, 19. 10. 1989, p. 1.

(^{1}) OJ No L 224, 18. 8. 1990, p. 29.
(^{2}) OJ No L 340, 11. 12. 1991, p. 17.
(^{3}) OJ No L 202, 14. 7. 1989, p. 14.
(^{4}) OJ No L 363, 24. 12. 1983, p.

2.5 Examples of amendments and modifications

Amendments/modifications appear in the form of Directives, Decisions or Regulations which clearly state in their title that it is an amendment or modification. For example, (Figure 2.4) 87/387/EEC has the title "Council Directive of 22 December 1986, amending Directive 75/318/EEC on the approximation of the laws of the member states relating to analytical, pharmaco-toxicological and clinical standards and protocols in respect of the testing of proprietary medicinal products". The actual changes (amendments) are clearly stated in the document. An example of a modification is shown in Figure 2.5.

Figure 2.4
Example of an EC Directive
which amends a previous
Directive

COUNCIL DIRECTIVE

of 22 December 1986

amending Directive 75/318/EEC on the approximation of the laws of the Member States relating to analytical, pharmaco-toxicological and clinical standards and protocols in respect of the testing of proprietary medicinal products

(87/19/EEC)

THE COUNCIL OF THE EUROPEAN COMMUNITIES,

Having regard to the Treaty establishing the European Economic Community, and in particular Article 100 thereof,

Having regard to the proposal from the Commission ([1]),

Having regard to the opinion of the European Parliament ([2]),

Having regard to the opinion of the Economic and Social Committee ([3]),

Whereas the testing of proprietary medicinal products must regularly be adapted to scientific and technical progress in order to ensure optimum protection of public health in the Community;

Whereas, in order to achieve such optimum protection of health, the resources allocated to pharmaceutical research must not be squandered on obsolete or repetitive tests resulting from divergences between the Member States in assessing the state of the art in science and technology;

Whereas, for ethical reasons, it is necessary to replace existing methods as soon as scientific and technical advances so allow by methods involving as few laboratory animals as possible;

Whereas, it is therefore necessary to introduce a rapid procedure for adapting to technical progress the requirements regarding the testing of proprietary medicinal products listed in the Annex to Directive 75/318/EEC([4]), as amended by Directive 83/570/EEC([5]), whilst ensuring close cooperation between the Commission and the Member States within a 'Committee for the Adaptation to Technical Progress of the Directives on the Removal of Technical Barriers to Trade in the Proprietary Medical Products Sector';

Whereas the requirements relating to the testing of medicinal products must also be capable of rapid revision by the same procedure, having regard to the evolution of test methods and of good laboratory practices recognized by the Community or in international trade in proprietary medicinal products,

HAS ADOPTED THIS DECISION

Article 1

Directive 75/318/EEC is hereby amended as follows:

1. The following Articles 2a, 2b and 2c shall be inserted:

([1]) OJ no C 293, 5. 11. 1984, p. 4.
([2]) OJ No C 36, 17. 2. 1986, p. 152.
([3]) OJ No C 160, 1. 7. 1985, p. 18.
([4]) OJ No L 147, 9. 6. 1975, p. 1.
([5]) OJ No L 332, 28. 11. 1983, p. 1.

Article 2a

Any changes which are necessary in order to adapt the Annex to take account of technical progress shall be adopted in accordance with the procedure laid down in Article 2c.

If appropriate, the Commission shall propose to the Council that the procedure in Article 2c be reviewed in connection with the detailed rules set for the exercise of the powers of implementation granted to the Commission.

Article 2b

1. A Committee on the Adaptation to Technical Progress of the Directives on the Removal of Technical Barriers to Trade in the Proprietary Medicinal Products Sector, hereinafter called "the Committee", is hereby set up; it shall consist of representative of the Member States with a representative of the Commission as chairman.

2.. The Committee shall adopt its own rules of procedure.

Article 2c

1. Where the procedure laid down in this Article is to be followed, matters shall be referred to the Committee by the chairman either on his own initiative or at the request of the representative of a Member State.

2.. The representative of the Commission shall submit to the Committee a draft of the measures to be adopted. The Committee shall deliver its opinion on the draft within a time limit set by the chairman, having regard to the urgency of the matter. It shall act by a qualified majority, the votes of the Member States being weighted as provided in Article 148 (2) of the Treaty. The chairman shall not vote.

3. (a) The Commission shall adopt the measures envisaged where they are in accordance with the opinion of the Committee.

(b) Where the measures envisaged are not in accordance with the opinion of the Committee, or if no opinion is adopted, the Commission shall without delay propose to the Council shall act by a qualified majority.

(c) If, within three months of the proposal being submitted to it, the Council has not acted, the proposed measures shall be adopted by the Commission.';

2. Part 1 of the Annex, 'Physico-Chemical, Biological or Microbiological Tests of Proprietary Medicinal Products', shall be amended as follows:

(a) In (A), the following section shall be inserted:

experimental studies validating the manufacturing process, where a non-standard method of manufacture is used or where it is critical for the products.';

(c) In (C) (2), subparagraph (b) shall be replaced by the following:

(b) the description of the substance, set down in a form similar to that used in a descriptive item in the European Pharmacopoeia, shall be accompanied by any necessary explanatory evidence, especially concerning the molecular structure where appropriate; it must be accompanied by an appropriate description of the method of synthetic preparation. Where substances can only be described by their method of preparation, the description should be sufficiently detailed to characterize a substance which is constant both in its composition and in its effects;'

3. Part 2 of the Annex, Toxicological and Pharmacological Tests'. is hereby amended as follows:

(a) The following paragraph shall be inserted after the introductory paragraph:

'Member States shall ensure that the safety tests are executed in conformity with the principles of good laboratory practice recognized by Community law in the field of tests on dangerous substances, or in the absence thereof, with those recommended by the Organisation for Economic Cooperation and Development.';

(b) In Chapter 1 (B), the text of paragraph 1 shall be replaced by the following:

'1. Single dose toxicity

An acute test infers a qualitative and quantitative study of the toxic reactions which may result from a single administration of the active substance of substances contained in the proprietary medicinal product, in the proportions and physico-chemical state in which they are present in the actual product.

The acute toxicity test must be carried out in two or more mammalian species of known strain unless a single species can be justified. At least two different routes of administration shall normally be used, one being identical with or similar to that proposed for use in human beings and the other ensuring systemic absorption of the substance.

This study will cover the signs observed, including local reactions. The period during which the test animals are observed shall be fixed by the investigator as being adequate to reveal tissue or organ damage or recovery, usually for a period of 14 days but not less than seven days, but without exposing the animals to prolonged suffering. Animals dying during the observation period should be subject to autopsy as also should all animals surviving to the end of the observation period.Histopathological examinations should be considered on any organ showing macroscopic changes at autopsy. The maximum amount of information should be obtained from the animals used in the study. The single dose toxicity tests should be conducted in such a way that signs of acute toxicity are revealed and the mode of death assessed as far as reasonably possible. In suitable species a quantitative evaluation of the approximate lethal dose and information on the dose effect relationship should be obtained, but a high level of precision is not required.

These studies may give some indication of the likely effects of acute overdosage in man and may be useful for the design of toxicity studies requiring repeated dosing on the suitable animal species.

In the case of active substances in combination, the study must be carried out in such a way as to check whether or not there is enhancement of toxicity or if novel toxic effects occur.'

Article 2

Member States shall take the measures necessary in order to comply with this Directive no later than 1 July 1987. They shall forthwith inform the Commission thereof.

Article 3

This Directive is addressed to the Member States.

Done at Brussels, 22 December 1986.

For the Council

The President

G. Shaw

Figure 2.5.
Example of an EC legislation
which modifies previous
legislation

COMMISSION REGULATION (EEC) No 762/92

of 27 March 1992

modifying Annex V to Council Regulation (EEC) No 2377/90 laying down a Community procedure for the establishment of maximum residue limits of veterinary medicinal products in foodstuffs of animal origin

THE COMMISSION OF THE EUROPEAN COMMUNITIES

Having regard to the Treaty establishing the European Economic Community,

Having regard to Council Regulation (EEC) No 2377/90 of 26 June 1990 laying down a Community procedure for the establishment of maximum residue limits of veterinary medicinal products in foodstuffs of animal origin[1], as amended by Commission Regulation (EEC) No 675/92 [2], and in particular Article 11 thereof,

Whereas it is desirable in the interests of administrative efficiency that the information and particulars to be included in an application for the establishment of a maximum residue limit for a pharmacologically active substance used in veterinary medicinal products in accordance with Regulation (EEC) No 2377/90 should correspond as closely as possible to the information and particulars to be submitted to Member States in an application for authorization to place a veterinary medicinal product on the market submitted in accordance with Article 5 of Council Directive 81/851/EEC of 28 September 1981 on the approximation of the laws of the Member States relating to veterinary medicinal products [3], as amended by Directive 90/676/EEC[4];

Whereas it is necessary to amend Annex V to Regulation (EEC) No 2377/90 to take account of the changes to the requirements for the testing of veterinary medicinal products introduced by Commission Directive 92/18/EEC of 20 March 1992 modifying the Annex to Council Directive 81/852/EEC

[1] OJ No L 224, 18. 8. 1990, p. 1.
[2] OJ No L 73, 19. 3. 1992, p. 8.
[3] OJ No L 317, 6. 11l 1981, p. 1.
[4] OJ No L 373, 31. 12

on the approximation of laws of the Member States relating to analytical, pharmacotoxicological and clinical standards and protocols in respect of the testing of veterinary medicinal products;

Whereas the provisions of this Regulation are in accordance with the opinion of the Committee on the Adaptation to Technical Progress of the Directives on the Removal of Technical Barriers to Trade in the Veterinary Medicinal Products Sector established under Article 2b of Council Directive 81/852/EEC [5], as amended by Directive 87/20/EEC[6],

HAS ADOPTED THIS REGULATION:

Article V to Regulation (EEC) No 2377/90 is hereby replaced by the Annex to this Regulation.

Article 2

This Regulation shall enter into force on the date of its publication in the Official

Journal of the European Communities.

This Regulation shall be binding in its entirety and directly applicable in all Member States. done at Brussels, 27 March 1992.

For the Commission

Martin BANGEMANN

Vice-President

[5] OJ No L 317, 6, 11. 1981 p16
[6] OJ No L 15, 17. 1. 1987 p34.

ANNEX

'ANNEX V

Information and particulars to be included in an application for the establishment of a maximum residue limit for a pharmacologically active substance used in veterinary medicinal products

Administrative particulars

1. Name or corporate name and permanent address of the applicant.

2. Name of the veterinary medicinal product

3. Qualitative and quantitative composition in terms of active principles, with mention of the international non-proprietary name recommended by the World Health Organisation, where such name exists.

4. Manufacturing authorization, if any.

5. Marketing authorization, if any.

6. Summary of the characteristics of the veterinary medicinal product(s) prepared in accordance with Article 5a of Directive 81/851/EEC.

A.   *Safety documentation*

A.0.   Expert report

A.1.   Precise identification of the substance concerned by the application

1.1   International non-proprietary name (INN).

1.2   International Union of Pure and Applied Chemistry (IUPAC) name.

1.3   Chemical Abstract Service (CAS) name.

1.4   Classification:

therapeutic;

pharmacological.

1.5   Synonyms and abbreviations.

1.6   Structural formula.

1.7   Molecular formula.

1.8   Molecular weight.

1.9   Degree of impurity

1.10   Qualitative and quantitative composition of impurities.

1.11 Description of physical properties:

melting point;

boiling point;

vapour pressure;

solubility in water and organic solvents, expressed in grams per litre, with indication of temperature;

density;

refractive index, rotation, etc.

A.2. Relevant pharmacological studies

2.1   Pharmacodynamics.

2.2   Parmacokinetics.

A.3. Toxicological studies

3.1   Single dose toxicity.

3.2   Repeated dose toxicity.

3.3   Tolerance in the target species of animal.

3.4   Reproductive toxicity, including teratogenicity.

3.4.1 Study of the effects on reproduction;

3.4.2 Embryotoxicity/Fetotoxicity, including teratogenicity.

3.5   Mutagenicity.

3.6   Carcinogenicity.

A.4. Studies of other effects

4.1   Immunotoxicity.

4.2   Microbiological properties of residues.

4.2.1 On the human fut flora;

4.3   Observations in humans.

B.   *Residue documentation*

B.0.   Expert report

B.1. Precise identification of the substance concerned by the application

The substance concerned should be identified in accordance with point A.1. However, where the application relates to one or more veterinary medicinal products, the product itself should be identified in detail, including:

- qualitative and quantitative composition;

- purity;

- identification of the manufacturer's batch used in the studies; relationship to the final product;

- specific activity and radio-purity of labelled substances;

- position of labelled atoms on the molecule.

B.2. Residue studies

2.1   Pharmacokinetics (absorption, distribution, biotransformation, excretion).

2.2   Depletion of residues.

2.3   Elaboration of maximum residue limits (MRLS).

B.3. Routine analytical method for the detection of residues

3.1   Description of the method.

3.2   Validation of the method.

3.2.1 specificity;

3.2.2 accuracy, including sensitivity;

3.2.3 precision;

3.2.4   limit of detection;

3.2.5 limit of quantitation;

3.2.6 practicability and applicability under normal laboratory conditions;

3.2.7 susceptibility to interference.'

2.6 Proposals

published in C
serie of OJ

A Proposal is exactly as the word says, the draft of a future piece of legislation. It will not become law until it is debated, negotiated, amended, re-written and issued as a Directive, Decision or Regulation. Proposals are published in the 'C' series of the OJ in order to give information (prior indication of things to come) or to allow feedback to the lawmakers (eg. through various lobby groups via EC Parliament).

2.7 How to find European Legislation relating to Biotechnology

- Electronic retreaval (CELEX)

CELEX (Communitatis Europeae Lex) is a computorized data base for EC law. It covers legislation, proposals, opinions, case law and parliamentary questions. It is an inter-institutional computerized documentation system for Community law, which has been open to the public since 1981. Starting from a document number, CELEX users can call up on their terminal the full texts of documents they wish to see. Moreover, they can see and obtain details of any amendments made to the piece of legislation in question between the time of the first publication and the present time. The data base is also issued on CD ROM (JUSTIS) which is up-dated every 6-months.

- The Official Journal of the European Communities

Most large city libraries and University libraries hold the OJ. Providing the user has the reference to an EEC document number and/an OJ number of the acts of interest, they can easily look up the relevant text of the legislation. Moreover, the OJ is issued with an index whereby the reader can look up keywords/expressions, descriptors and non-descriptors in an Alphabetical table. Thus, legislation relating to a given topic (eg. "biotechnology") can be sought by looking up this keyword in the table and obtaining EEC document numbers and/or OJ reference numbers which correspond with this topic. It is then possible to obtain the full text from the OJ. However, it is not possible to find out from the OJ if later amendments have been made to any particular pieces of legislation. For this a different reference source is needed, the various "Directories".

- The "Directory of Community Legislation in Force"

The EC office of Official Publications publishes a Directory which is recompiled and updated every 6 months. It gives references to all binding Community legislation in force at the time of publication. It therefore includes (i) agreements and conventions concluded by the Communities in connection with their external relations with non-member countries; (ii) binding secondary legislation (Regulations, Decisions, ECSC general decisions and recommendations, EEC/Euratom Directives) under the Treaties establishing the European Communities (with the exception of day-to-day administrative acts); (iii) supplementary legislation, in particular decisions of representatives of the Governments of the Member States meeting within the Council; (iv) certain non-binding acts (opinions, proposals) considered by the institutions to be important. The Directory appears in two volumes; volume I is the main body of the directory and consists of 17 chapters with acts arranged according to subject. For example, chapter 15 is "Environment, consumers and health protection" and contains references to many of the acts which relate to biotechnology. Volume II contains both chronological and alphabetical indexes of the acts appearing in volume I. Fortunately, those unfamiliar with the Directory can read the section in volume I entitled "Information for Readers" which explains to readers how to use the Directory and gives an explanation of codes. The great advantage of the OJ Directory is that following a particular entry number and the title of the act in question is given a list of subsequent amendments and supplementary acts.

- Other Directories:

  1). The Annual General Report on the Activities of the European Communities, published by the Commission. Biotechnology is mentioned as a heading under various sections (eg. in the Research & Technology section).

  2). The Monthly EC Bulletin. This is similar to the annual report except that it is issued monthly.

  3). European Current Law (published by Sweet & Maxwell)

Biotechnology and the law: summary of some current legislation in force

Biotechnology and the law: summary of some current legislation in force

3.1 Introduction

This chapter gives an overview of some of the current "regulatory" legislation in force which relates to biotechnology. It is generally true that products of traditional manufacturing processes are subject to regulations (eg. on product safety) whereas the manufacturing process itself is relatively free of specific regulatory controls. However, biotechnology is an important exception to this principle since regulatory provisions exist at every stage of the development of biotechnology services or products. For example, at the research stage, controls exist to help ensure the safety and well-being of laboratory workers including controls on the manipulation and release of genetically engineered organisms. Legislation covers good laboratory practice and animal experimentation. At the production stage there are important laws on the health and safety of technical staff and other workers as well as on product testing and analysis. Finally, there are important laws which govern all processes which may cause emissions of harmful or polluting substances. These laws (covering environmental pollution) are widely applied to all industrial processes, but are of particular importance to those using new technologies (eg. environmental biotechnology) in order to treat toxic compounds and reduce emissions of pollutants at source. Many of the topics covered in this chapter will be expanded upon in later chapters.

range of legislation

Implementation of EC legislation through national legislation

It is the nature of EC Directives that they have to be translated into the national legislation of the various member states. Thus, national legislation is becoming little more than a reflection of the wider legal framework established by the EC. This chapter therefore tends to adopt primarily an EC perspective of legislative controls. However, on certain topics central to biotechnology separate national legislation exist. It is imperative therefore that biotechnology is conducted within the conditions specified by EC and national authorities. It is important that you are aware of obligations at both EC and national levels. It is beyond the scope of this text to deal with all the separate national regulations although, in some instances, we will cite specific examples.

Most of the EC legislation which relates to biotechnology is in the form of Directives (see chapters 1 and 2). Although Directives set out the main principles and aims of legislation, they leave the implementation to be undertaken by the Member States, usually within a specified time limit. In the UK, for example, the implementation of an EC Directive can be done in one of four different ways:

New primary legislation can be drafted which takes into account the Directive (eg. Consumer Protection Act, 1987).

Secondary or delegated legislation can be enacted where the primary legislation such as the Health and Safety at Work etc. Act 1974 provides adequate power to make delegated legislation (ie. later "additions") by Statutory Instruments such as the Control of Substances Hazardous to Health Regulations 1988 (COSHH

Regulations). The new Instruments can therefore encompass the various EC Directives. It should be noted that Regulations are laid before Parliament by the relevant Secretary of State and can be rapidly adopted whereas Acts require detailed parliamentary debate and normally take considerable time.

The European Communities Act 1972 includes general powers to implement any Directive by delegated legislation and can be used where sector-specific primary legislation is not adequate or apt, or does not have provisions to adopt the EC Directive in question.

Directives may be implemented by UK Administrative Action (for example, compliance with EC Directive 87/18/EEC on Good Laboratory Practice). This latter method is not always considered adequate by the Commission since it is more difficult for the Commission to ascertain whether or not the original aims of the Directive have been fully implemented.

Other Member States of the EC have analogous procedures for the implementation of EC Directives although each differs in detail in terms of precise mechanisms. Nevertheless, you must be aware that EC Directives are ultimately incorporated in some form or other into national legislation.

In this overview of biotechnology and the law, we have divided discussion into:

- health and safety of workers;
- good laboratory practice (GLP);
- use of animals in experiments;
- propriety medical and veterinary products;
- environmental pollution;
- patenting.

Food production and processing are also subjects of many EC-Directives. Many of these deal with such issues as labelling and market restrictions. (see Chapter 12)

We have represented these areas diagrammatically in Figure 3.1 and listed some of the important EC-Directives. We have included the titles of these Directives, their purposes and publication references in Table 3.1. There is a lot of information in Table 3.1 but it is well worth reading through this table at this stage as it will give you an overview of the issues dealt with by the various Directives. You should also refer to this Figure and Table as you read through the rest of this chapter. Many of the Directives and other measures will be discussed in further detail in the subsequent chapters.

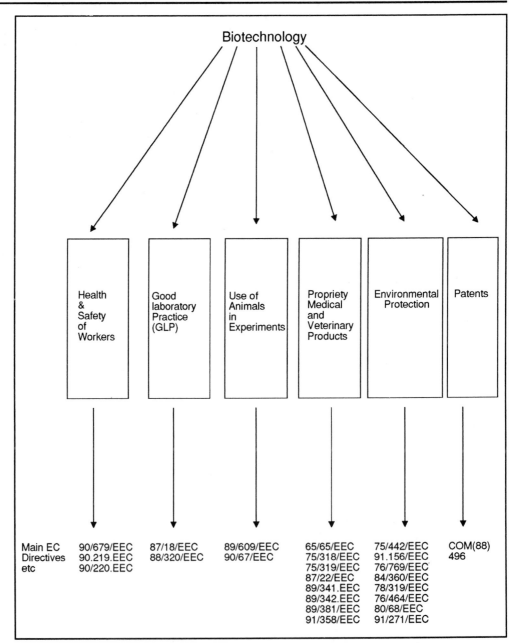

Figure 3.1 The main EC-Directives and Decisions relating to important aspects of Biotechnology. Note that we have not attempted to cover all the relevant Directives here. There are many additional Directives which have wide application which also apply to Biotechnological enterprises. Some for example relate to worker safety, some to financial practices and so on. A brief summary of some of the Directives cited above are included in Table 3.1. In Table 3.1 we have used a slightly different arrangement of sections. Note also that EC-Directive and Decisions relating to food and food processing have not been included in Figure 3.1 nor Table 3.1. since most of these do not relate to biotechnological activities. Examples and comments on the Directives relating to food are given Chapter 12.

TOPIC AND TITLE OF LEGISLATION (LEGISLATION REF. NO.)	PURPOSE	DATE BROUGHT INTO FORCE	PUBLICATION REFERENCE
HEALTH & SAFETY			
Directive on the protection of workers from risks related to exposure to biological agents at work (90/679/EEC)	To protect workers against risks to their health and safety, including theprevention of such risks, arising or likely to arise fromexposure to biological agents at work.	29.11.93	OJ L374 31.12.90
Directive on the contained use of genetically modified organisms(90/219/EEC)	To establish common measures for the contained use of GMO's with a view to protecting human helath and theenvironment	23.10.91	OJ L117 08.05.90
Directive on the deliberate release of genetically modifiedorganisms into the environment (90/220/EEC)	To protect human health and the environment when carrying out deliberate releaseof GMO's or when placing on the market products containing/consisting of GMO's for subsequent release	23.10.91	OJ L117 08.05.90
GOOD LABORATORY PRACTICE AND THE USE OF ANIMALS			
Directive on the harmonisation of laws, regulations and administrative provisions relating to the application of the principles of good laboratory practice and the verification of their applications for tests on chemical substances (87/18/EEC)	To ensure that laboratories carrying out tests on chemical products (in accordance with Directive 67/548/EEC) comply with the principles of good laboratory practice. To ensure conformity with the principles of good laboratory practice in respect of tests on chemical products to evaluate their safety for man and/or the environment.	30.6.88	OJ L015 17.01.87
Directive on the inspection and verification of Good Laboratory Practice (GLP) (88/320/EEC)	To ensure that inspection and verification of laboratories and laboratory procedures (planning,, testing, recording, reporting) are carried out in accordance with the rules and regulations in force and in compliance with GLP.(Directive is not concerned with the interpretation and evaluation of test results; only the practice procedures)	01.01.89	OJ L145 11.06.88
Directive on the approximation of laws, regulations and administrative provisions of the Member States regarding the protection of animals used for experimental and other scientific purposes (86/609/EEC	To ensure that where animals are used for experimental or other scientific purposes, the provisions laid down by law, regulation or administrative provisions in the Member States for their protection are approximated so as to avoid affecting the establishment and functioning of the common market, in particular by distortions of competition or barriers to trade.	24.11.89	OJ L358 18.12.86
Decision setting up an advisory committee on the protection of animals used for experimental and other scientific purposes (90/67/EEC)	A permanent consultative committee (all Member States represented) of experts in the field of animal experimentation who have experience of regulations and administrative practices who can assist the Commission (can respond effectively to questions and help exchange information). Related to Directive 86/609/EEC.	09.02.90	OJ L044 20.02.90
PROPRIETARY MEDICAL OR VETERINARY PRODUCTS			
Proposal for a Council Directive on the legal protection of biotechnological inventions (COM(88)496)	Adaptation of European patent system to new technological developments including the patenting of living matter. To ensure approximated protection throughout the Member States.	Proposal only	OJ C010 13.01.89
Directive on the approximation of provisions laid down by law regulation or administrative action relating to proprietary medicinal products (65/65/EEC)	To remove disparities between national provisions relatingto medicinal products in order to help in the establishment andfunctioning of the common market.	26.06.66	OJ L22 09.02.65

Table 3.1 Summary of some important EC-Directives which have major importance in Biotechnology (continued)

Directive on the approximation of the laws of the Member States relating to analytical, pharmacotoxicological and clinical standards and protocols in respect of the testing of proprietary medicinal products. (75/318/EEC)	To ensure uniformity of standards, protocols, trials and tests in the evaluation of efficacy and harmfulness of medicinal products; to avoid differences in the evaluation of medicinal products by adopting common procedures. (Annexes show requirements).	20.05.75	OJ L147 09.06.75
Second Council Directive on the approximation of provisions laid down by law, regulation or administrative action relating to proprietary medicinal products (75/319/EEC).	To extend and implement directive 65/65/EEC. and set up a committee consisting of representatives of Member States and the Commission (the Committee for Proprietary Medicinal Products)	20.11.76	OJ L147 09.06.75
Directive on the approximation of national measures relating to the placing on the market of high-technology medicinal products, particularly those derived from biotechnology (87/22/EEC).	To ensure identical rules thoroughout the EC governing the placement on the market of high-technology medicinal products; to extend previous directives on approximation of laws to products of biotechnology (inc. rDNA products and monoclonal antibodies)	01.07.87	OJ L1517.01.87
Directive amending Directives 65/65/EEC, 75/318/EEC on the approximation of provisions laid down by law, regulation or administrative action relating to proprietary medicinal products. (89/341/EEC)	Amends previous directives in order to cover other industrially produced medicinal products that were hitherto excluded.	01.01.92	OJ L142 25.05.89
Directive extending the scope of Directives 65/65/EEC and 75/319/EEC and laying down additional provisions for immunological medicinal products consisting of vaccines, toxins or serums and allergens, (89/342/EEC).	Extends previous directives in order to include vaccines, toxins, serums and allergen products. Ensures that product composition is expressed in terms of biological activity or protein content.	01.01.92	OJ L142 25.05.89
Directive extending the scope of Directives 65/65/EEC and 75/319/EEC on the approximation of provisions laid down by law,, regulation or administrative action relating to proprietary medicinal products and laying down special provisions for medicinal products derived from human blood or human plasma (89/381/EEC).	Extends previous directives to include medicinal products derived from human blood or human plasma, in particular,, albumin, coagulating factors and immunoglobulins.	01.01.92	OJ L181 28.06.89
Directive laying down the principles and guidelines of good manufacturing practice for medicinal products of human use (91/356/EEC)	To help ensure that all medicinal products for human use (imported or manufactured) should be produced in accordance with GLP, to help with implementation of pharmaceutical quality assurance and reporting systems. To help with approximation of laws as outlined in Directive 75/318/EEC.	02.07.91	OJ L193 17.07.91
POLLUTION AND WASTE			
Directive on waste (75/442/EEC)	To encourage the prevention and recycling of waste. To determine the arrangements to be made for the harmless disposal of waste. To provide administrative provisions for management and control.	18.07.77	OJ L194 25.07.75

Table 3.1 Summary of some imporatant EC-Directives which have major importance in Biotechnology. (continued)

Directive amending 75/442/EEC on waste (91/156/EEC)	Tightens up definition of wastes and lists 16 specific categories. Obligations of the 1975 Directive are expanded.	25.03.91	OJ L78 26.03.91
Directive relating to restrictions on the marketing and use of certain dangerous substances and preparations (76/769/EEC)	To create a general framework for restrictions on the marketing and use of dangerous substances (PCB's, PCT's, monomer vinyl chloride)	03.02.78	OJ L262 27.09.76
"Framework" Directive on combating air pollution from industrial plants (84/360/EEC)	To implement a system of measures and procedures which will prevent and reduce air pollution from industries within the EC.	30.06.87	OJ L188 16.07.84
Directive on toxic and dangerous wastes (78/319/EEC)	To promote the prevention and recycling of toxic and dangerous wastes. To lay down the arrangements to be made for its safe disposal. To provide administrative measures for management and control (inc. a system of authorisation for firms to store, treat or deposit toxic/dangerous waste. To provide for disposal programmes (which shall be notified to the Commission) and to provide for action to be undertaken in emergencies. To record disposal of toxic and dangerous wastes	22.03.80	OJ L84 31.03.78
Directive on the protection of the environment, and in particular of the soil, when sewage sludge is used in agriculture.	To provide for a special regime concerning the spreading of sludge in agriculture and to fix sludge and soil analyses.	17.06.89	OJ L181 04.07.86
Directive on pollution caused by certain dangerous substances discharged into the aquatic environment of the Community (76/464/EEC)	To provide a system of authorisations for the discharge of dangerous substances into water. To provide limitvalues or quality objectives and monitoring procedures for List I substances. To provide quality objectives to List II substances. To adopt anti-pollution programmes for both types of substances and communicate them to the Commission. To draw up a list of discharges involving List I substances.	04.05.78	OJ L129 18.05.76

Table 3.1 Summary of some important EC-Directives which have major impact on Biotechnology.

3.2 Health and safety of workers

3.2.1 Key Directives relating to genetic modification

production and release of genetically modified organisms

There are two EC-Directives relating to genetically modified organisms. EC Directive 90/219/EEC provides the framework for safety of workers who are employed to use genetically modified micro-organisms in a, so called, contained manner. The EC Directive 90/220/EEC describes the procedures and conditions under which genetically modified organisms may be deliberately released.

These two Directives are so central to biotechnology that we have dedicated a whole chapter to them so we will not discuss then in greater detail here. We point out, however, that these Directives are leading to, or have led to, changes to existing national legislation or have been the basis for the formation of new legislation within Member States. We will illustrate this by using specific examples.

In the UK, The Health and Safety of Work Act of 1974 is the main piece of legislation aimed at ensuring the safety of employers. Most of the controls relating to biotechnology were first established through this Act. Within the framework of this Act, the Health and Safety (Genetic Manipulation) Regulations were introduced in 1978. These Regulations have now been revoked and replaced by the Genetic Manipulation Regulations 1989. These later Regulations were designed to fulfil the requirements of the EC Directives 90/219/EEC and 90/220/EEC. In this case we see an example of changes to national regulations mediated by EC Directives. A similar adjustment to national regulations relating to the release of genetically modified organisms (GMO's) has also been made by the introduction of provisions in the Environmental Protection Act 1990.

We can therefore see a progressive harmonisation of national legislation within the EC. We have illustrated this process in Figure 3.2.

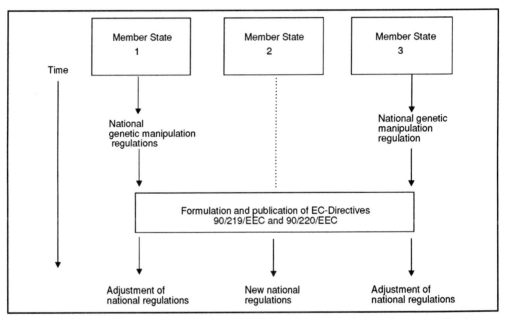

Figure 3.2 Interplay between national regulations and EC-Directives relating to genetic manipulation (see text for discussion.

In the situation shown in Figure 3.2. Member States 1 and 3 had existing regulations prior to the production of the EC-Directives. They subsequently have adjusted their regulations to fulfil the requirements of the Directives. Member State 2 on the other hand, had no existing legislation but subsequently produced regulations to fulfil obligations arising from the EC-Directives. Although the route by which national regulations are produced and implemented differ in different Member States, the provisions of such regulations fulfil the same criteria. This example also illustrates why awareness of the provisions and contents of EC-Directives is important. Knowledge of the relevant EC-Directives provides the basis for understanding the provisions of regulations in other Member State as well as the regulations of your home country.

3.2.2 Key Directives relating to Substances Hazardous to Health.

COSHH

There are many EEC Directives relating to the use of substances hazardous to health. The main conditions and procedures are specified within the EC Directives 89/391/EEC, 90/679/EEC, 90/394/EEC. These conditions have also been incorporated into national legislation. For example in the UK, the Control of Substances Hazardous to Health (COSHH) Regulations have also been made under the Health and Safety at Work Act 1974. The key Statutory Instruments are S1 1988/1657 and SI 1990/2026. The COSHH Regulations cover all chemical substances used in research and manufacture (There are a few cases which are exempt, for example some medicinal products which are used to treat patients).

It is vital therefore that you are aware of the Regulations operative in your geographical region. Although some national differences exist in the way the various EC Directives have been interpreted the underpinning principles are similar. (see also Chapter 13) The aim of these measures is to protect the health and safety of workers. The way they achieve this is by:

- requiring production of an inventory of all the chemical substances used in the process;

- requiring identification of hazardous substances;

- demanding that an assessment of risks from substances is made;

- demanding use of control measures to minimise risks, including:

    substitution;

    control of the process;

    ventilation;

    respiratory protection/protective clothing.

- requiring careful monitoring of:

    personal exposure;

    the control measures;

    health effects;

- requiring records of monitoring;

- requiring reassessment if significant changes are made in the process.

3.2.3 Radiation

Radioactive isotopes are used extensively in Biotechnology as labels to trace and quantify materials. Such isotopes are potentially dangerous and present other hazards than chemical reactions. We will examine radiation and the regulations governing its use in Chapter 7.

3.3 Good Laboratory Practice

role of FDA and OECD

In the US, the concept of Good Laboratory Practice (GLP) was introduced in order that high standards of laboratory testing could be maintained across the nation and that the Food and Drug Administration (FDA) and other regulatory and registration authorities could rely with some degree of confidence that the test data obtained about a new drug or chemical were reliable. The Council of the Organisation for Economic Cooperation and Development (OECD) took a Decision on 12 May 1981 on the mutual acceptance of data for the evaluation of chemical products. A later OECD recommendation was made in July 1983 concerning the mutual recognition of compliance with GLP. Three EC Council Directives (87/18/EEC, 88/320 EEC and 90/18/EEC) were issued in order to harmonise the laws, regulations and administrative provisions relating to the applications of the principles of GLP and the verification of their application for tests on chemical substances.

The EEC were sensible in adopting legislation which would harmonise laws on chemical testing and GLP since it would ensure that EC Member States would recognise each others' test results on chemicals and cut out the need to duplicate tests and cause unnecessary wastage of resources. In addition, the legislation would reduce considerably the numbers of experiments carried out on animals where toxicity data were needed. Clearly, in order to operate such a system it requires that tests are carried out using standard, well recognised procedures and that the test data are reliable and recognised and accepted by all member states. In order to be confident about this, a harmonised system for study audit and inspection of EC laboratories would also be required in order to ensure that all laboratories were working to the same standards of GLP conditions. In the UK, GLP is dealt with by a GLP Compliance Monitoring Unit at the Department of Health. This administrative unit deals with all the necessary GLP certification, study audit and inspection of laboratories. Thus, in the UK the three EC Directives are implemented by the administrative authorities alone and (at the time of writing) no specific UK laws have been drafted in this area.

3.4 Use of Animals in Experiments

Many countries have had long standing legislation regarding the use (abuse) of animals. For example, in the UK all animal experimentation was regulated by the Cruelty to Animals Act 1876. However the EC has formulated standards (86/609/EEC) which has led to a harmonisation on regulations regarding the use of animals for experimental purposes. For example, in the UK, the 1876 act has been replaced by the Animals (Scientific Procedures) Act 1986.

Because of the importance of EC-Directive 86/609/EEC in biotechnology, we have included a full chapter (Chapter 7) on this Directive.

3.5 Proprietary Medical and Veterinary Products

A substantial number of EC Directives have been issued over the last 25 years in an attempt to harmonise the various laws operating in each of the Member States on proprietary medicinal products. Clearly, it is important in establishing a single European Market that large disparities in the rules relating to medicinal products are removed. This process is not easy since the rules which pertain to medicinal compounds are usually extremely stringent. Here we provide some outline guidance since we will provide greater details in chapters 10 and 11.

Medicinal products are defined in Directive 65/65/EEC as substances used for treating/preventing diseases in either human beings or animals and as a term it therefore covers both veterinary and human medicines. The following EC Directives are concerned with approximation (bringing into line) the various laws of the different Member States: 65/65/EEC, 75/318/EEC, 75/319/EEC, 89/341/EEC, 89/342/EEC, 89/343/EEC and 89/381/EEC. Basically, these laws are harmonisation measures which aim to standardise the tests, standards, protocols, regulations and administrative systems which relate to medicinal products. Directive 75/319/EEC provided for the establishment of The Committee for Proprietary Medicinal Products (CPMP) to help in the establishment of a common procedure for obtaining market authorizations and in establishing a monitoring procedure. (NB. A "market authorization" under EC Directives is equivalent to a UK "product licence"). EC Directives have also been made in related areas such as approximation of the laws of Member States relating to colouring matters which may be added to medicinal products (78/25/EEC and 81/464/EEC). Some EC legislation (eg. 83/570/EEC, 87/19/EEC and 87/21/EEC) are amendments of earlier Directives concerned with approximation whilst others (eg. 87/18/EEC, 88/320/EEC and 90/18/EEC) are concerned with the application and verification of Good Laboratory Practice.

CPMP and market authorisation

Of special interest to biotechnologists are the following Directives: 87/22/EEC (on the approximation of national measures relating to the placing on the market of high-technology medicinal products, particularly those derived from biotechnology), 89/342/EEC (extending the scope of Directives 65/65/EEC and 75/319/EEC and laying down additional provisions for immunological medicinal products consisting of vaccines, toxins or serums and allergens), and 89/381/EEC (also extending the scope of Directives 65/65/EEC and 75/319/EEC, but laying down additional provisions for medicinal products derived from human blood or plasma such as albumin, coagulating factors and immunoglobulin).

Some Community measures are specific to veterinary medicinal products (eg. 81/851/EEC, 81/852/EEC and 87/20/EEC) but many of the recent Directives (eg the extension of legislation to high technology medicines particularly those derived from biotechnology, 87/22/EEC) apply to both human and veterinary products.

product licence

Most EC laws to date on medicinal compounds are concerned with "Market Authorization" which is equivalent to "Products Licence" under the UK Medicines Act 1968. At the time of writing, no EC legislation has been drafted in the area of clinical trials testing. It is in this respect that UK law is in advance of EEC law since under the Medicines Act 1968, regulations have been added (Medicines [Standard Provisions for Licences and Certificates] Regulations 1971; SI 1971/972 as last amended by SI 1983/1730) which ensure that phase I, II and III clinical trials are carried out for all new products in an ethical manner to ensure safety and efficacy of the medicinal product in question.

3.6 Environmental Pollution

Environmental pollution is becoming a major issue and prevention of pollution is becoming the subject of much legislation. Pollution can be in many forms; biological, chemical, thermal, nuclear, noise etc. Many EC Directives have been published relating to this issue. Most of these are generally applicable and are not specific to biotechnology so we will not examine them in depth here. The EC Directive 90/220/EEC relating to the deliberate release of genetically modified organisms described earlier is essentially an environmental protection measure

3.6.1 Chemical substances

Black and Grey
list substances

The classification of dangerous substances and laws relating to the control of their discharges are mainly dealt with through EEC Directives 76/464/EEC; 80/68/EEC; 76/769/EEC; and 84/360/EEC. The overall aim of the Directives are to protect the environment against pollution and to reduce the risks to human and other forms of life. The main Directive (76/464/EEC) describes two classes or groups of toxic substances. The first group are "List I" chemicals (often referred to as the "Black List") which are considered to be very toxic and likely to persist in the environment. Examples include mercury, cadmium, organohalides (eg PCB's), organophosphates, the "drin-pesticides" (aldrin, endrin, dieldrin, isodrin). The second group ("List II", the "Grey List") are also toxic chemicals, but whose presence is considered to be less harmful when discharged. These include various biocides, cyanide, ammonia, zinc, nickel, chromium, lead, arsenic and copper. In the UK, the Environmental Protection Act 1990 (EPA) was introduced in order to protect the environment (air, land and water) from pollution. It is mainly through this act that the various EC Directives on the environment are implemented in the UK.

Similar legislations have been introduced in other EC-Member States. In the Netherlands, for example, the various EC Directives on the environment are implemented mainly through the Wet miliengevaarlyke stoffen (WMS), 1985 ("Hazardous chemical compounds Act") and the Bestrijdingsmiddelen wet, 1975 ("Pesticide Act".)

3.6.2 Discharges to air, water and land and integrated pollution control

National regulations and systems of enforcement show much variation. There is, however, an increasing body of EC-Directives which deals with such issues. The Air Framework Directive 84/360/EEC on the Combatting of Air Pollution from Industrial Plants is a key Directive. This Directive introduced the term "BATNEEC" (best available technique not entailing excessive cost). The principle of BATNEEC has also been applied to discharges to water and land.

- "Best" means the most effective in preventing, minimising or rendering harmless polluting emissions.

- "Available" means generally accessible and procurable by the operator. It does not imply that the technology is in general use, but it does require general accessibility.

- For new processes, where the costs of best available techniques would be excessive "not entailing excessive cost" means that proposals can be modified by economic considerations. For existing processes, timescales will be established within which old processes will have to be upgraded to new standards or decommissioned.

Many other catchy phrases have been adopted by Member States For except, "Best Practicable Environmental Option (BPEO) was introduced in the UK in 1975 15th Report of the Royal Commission on Environmental Pollution. BPEO has been defined as.

> "The optional allocation of the waste especially the use of different sectors of the environment to minimise damage overall" (12th Report of the Royal Commission on Environmental Pollution 1988).

The protection of water against deterioration of water quality and pollution is the subject of a wide range of EC-Directives (see Table 3.2.)

Directive number	Topic coverage
76/464/EEC	Framework Directive - Black and Grey list substances
80/68/EEC	Protection of ground water
75/440/EEC	Surface water quality
76/160/EEC	Bathing water quality
80/778/EEC	Drinking water quality

Table 3.2 Main EC-Directives relating to the protection of waters

pollution control

Increasingly emphasis is being placed on an integrated approach to pollution control. In other words design and operation of processes are considered together.

It must be pointed out that biotechnological processes must, like any other process, meet the conditions established by such Directives and associated regulations. The importance of this legislation to biotechnolgy is not only one of controlling the way in which biotechnology operates, but also provides many opportunities for exploiting biotechnology. Biotechnology is, in essence, a clean technology and provides opportunities to replace other, dirtier technologies in the manufacture of some products. Biotechnology also provides opportunities to develop "end-of-pipe" processes to remove or transform toxic materials discharged from some conventional processes.

3.7 Protection of invention - patenting

So far our discussion about Biotechnology and law has focused onto issues relating to what we might term quality assurance, good practice and the protection of workers and the environment. Another important aspect of legislation in relation to biotechnology relates to mechanisms that protect the interests of inventors and those who invest in research and development. Inventions may be protected by patent(s). The patenting system was not, however, particularly designed to cope with processes based on biological materials and activities and this raised great concern about the value of the patent system to protect biotechnological inventions. This in turn threatened investment in this area. However a proposal for a Council Directive on the legal

protection of biotechnological invention

protection of biotechnological inventions has been made (COMM (88) 496). The aim of this new directive is to adapt the European patent system to new technological developments including the patenting of living matter to ensure protection throughout Member States.

The issues of patenting are discussed more fully in Chapter 9.

Good laboratory practice

Good laboratory practice

4.1 Introduction

Many countries have passed legislation to control the production and use of chemicals. This legislation usually requires the manufacturer to perform laboratory studies and to submit the results of these studies to governmental authorities for assessment of the potential hazards to human health. For a long time, much concern was expressed over the quality of studies upon which hazard assessments were based. As a consequence, several states, especially members of OECD, jointly planned to establish criteria for the performance of these studies. These OECD Member countries recognised that it was important to harmonise test methods and laboratory practices to facilitate international acceptability and trade.

An international group of experts established under a Special Programme on the Control of Chemicals, developed a document concerning the 'Principles of Good Laboratory Practice (GLP)'. This utilised common managerial and scientific practices and experience from a wide range of national and international sources.

The OECD's document has provided the framework in which subsequent regulations have been formulated. Within the EC, two Directives are particularly relevant. They are 87/18/EEC and 88/320/EEC

EC Directive 87/18/EEC is the 'Directive on the harmonisation of laws, regulations and administrative provisions relating to the application of the principles of good laboratory practice and the verification of their application for tests on chemical substances". (OJ L015 17.01.87)

The purpose of this Directive is to ensure that laboratories carrying out tests on chemical products (in accordance with Directive 67/548/EEC) comply with the principles of Good Laboratory Practice. The aim is to ensure conformity with the principles of Good Laboratory Practice in respect of tests on chemicals to evaluate their safety for Man and/or the Environment.

EC Directive 88/320/EEC is the "Directive on the inspection and verification of Good Laboratory Practice (GLP)" (OJ L145 11.06.88)

The purpose of this Directive is to ensure that inspection and verification of laboratories and laboratory procedures (including planning, testing, recording, reporting) are carried out in compliance with GLP. The Directive is not concerned with the interpretation and evaluation of test results but only with procedures.

Here we will summarise the principles of Good Laboratory Practice in the assessment of the potential hazards to human health arising from chemical substances.

4.2 Summary of 'principles of good laboratory practice (GLP)' established by OECD members

Below is a list of issues covered by GLP. The numbers in brackets indicate the sections in which we will examine a specific aspect of GLP. We have chosen not to attempt to cover all of the sections.

Scope (Section 4.3)

Definition of terms (Section 4.4)

Good Laboratory Practice (4.4.1)

Terms Concerning the Organisation of a Test Facility (4.4.2)

Terms Concerning the Study (4.4.3)

Terms Concerning the Test Substance (4.4.4)

Good Laboratory Practice Principles (Section 4.5)

Test Facility Organisation and Personnel (4.5.1.)

Management's Responsibilities

Study Director's Responsibilities

Personnel Responsibilities

Quality Assurance Programme (4.5.2)

General

Responsibilities of the Quality Assurance Personnel (4.5.3)

Facilities

General

Test System Facilities

Facilities for Handling Test and Reference Substances

Archive Facilities

Waste Disposal

Apparatus, Material, and Reagents (4.5.4)

Apparatus

Material

Reagents

Test Systems (4.5.5)

Physical/Chemical

Biological

Test and Reference Substances (4.5.6)

Receipt, Handling, Sampling, and Storage

Characterisation

Standard Operating Procedures (4.5.7)

General

Application

Performance of the Study (4.5.8)

Study Plan

Content of the Study Plan

Conduct of the Study

Reporting of Study Results (4.5.9)

General

Content of the Final Report

Storage and retention of Records and Material (4.5.10)

Storage and Retrieval

Retention

We have reported on each of these in note form to provide a succinct checklist so that you are aware of the issues involved in GLP. You would, of course need to consult the definitive documents(s) for the details of each of these points.

4.3 Scope

These Principles of Good Laboratory Practice should be applied to the testing of chemicals to obtain data on their properties and/or their safety with respect to human health or the environment.

Studies covered by Good Laboratory Practice also include work conducted in field studies. These data would be developed for the purpose of meeting regulatory requirements.

4.4 Definitions of Terms

4.4.1 Good Laboratory Practice (GLP)

GLP is concerned with the organisational process and the conditions under which laboratory studies are planned, performed, monitored, recorded and reported

4.4.2 Terms concerning the organisation of a test facility

- **Test facility** means the persons, premises, and operational unit(s) that are necessary for conducting the study.

- **Study Director** means the individual responsible for the overall conduct of the study.

- **Quality Assurance Programme** means an internal control system designed to ascertain that the study is in compliance with these Principles of Good Laboratory Practice.

- **Standard Operating Procedures (SOPs)** means written procedures which describe how to perform certain routine laboratory tests or activities normally not specified in detail in study plans or test guidelines.

- **Sponsor** means a person(s) or entity who commissions and/or supports a study.

4.4.3 Terms concerning the study

- **Study** means an experiment or a set of experiments in which a test substance is examined to obtain data on its properties and/or its safety with respect to human health and the environment.

- **Study plan** means a document which defines the entire scope of the study.

- **OECD Test Guideline** means a test guideline which the OECD has recommended for use in its Member countries.

- **Test system** means any animal, plant, microbial, as well as other cellular, sub-cellular, chemical, or physical system or a combination thereof used in a study.

- **Raw data** means all original laboratory records and documentation or verified copies.

- **Specimen** means any material derived from a test system for examination, analysis, or storage.

4.4.4 Terms Concerning the Test Substance

- **Test substance** means a chemical substance or a mixture which is under investigation.

- **Reference substance (control substance)** means any well defined chemical substance or any mixture other than the test substance used to provide a basis for comparison with the test substance.

- **Batch** means a specific quantity or lot of a test or reference substance produced during a defined cycle of manufacture in such a way that it could be expected to be of a uniform character and should be designated as such.

- **Vehicle (carrier)** means any agent which serves as a carrier used to mix, disperse, or solubilise the test or reference substance to facilitate the administration to the test system.

- **Sample** means any quantity of the test or reference substance.

4.5 Good Laboratory Practice Principles

4.5.1 Test facility organisation and personnel

Management's Responsibilities

Test facility management should ensure that the Principles of Good Laboratory Practice are complied within the test facility.

At a minimum it should:

   a) ensure that qualified personnel, appropriate facilities, equipment, and materials are available;

b) maintain a record of the qualifications, training, experience and job description for each professional and technical individual;

c) ensure that personnel clearly understand the functions they are to perform and, where necessary, provide training for these functions;

d) ensure that health and safety precautions are applied according to national and/or international regulations;

e) ensure that appropriate Standard Operating Procedures are established and followed;

f) ensure that there is a Quality Assurance Programme with designated personnel;

g) where appropriate, agree to the study plan in conjunction with the sponsor;

h) ensure that amendments to the study plan are agreed upon and documented;

i) maintain copies of all study plans;

j) maintain a historical file of all Standard Operating Procedures;

k) for each study ensure that a sufficient number of personnel is available for its timely and proper conduct;

l) for each study designate an individual with the appropriate qualifications, training, and experience as the Study Director before the study is initiated. If it is necessary to replace a Study Director during a study, this should be documented;

m) ensure that an individual is identified as responsible for the management of the archives.

Study Director's Responsibilities

These responsibilities should include, but not be limited to, the following functions:

a) ensure that the procedures specified in the study plan are followed, and that authorization for any modification is obtained and documented together with the reasons for them;

b) ensure that all data generated are fully documented and recorded;

c) sign and date the final report to indicate acceptance of responsibility for the validity of the data and to confirm compliance with these Principles of Good Laboratory Practice;

d) ensure that, after termination of the study, the study plan, the final report, raw data and supporting material are transferred to the archives.

Personnel Responsibilities

a) Personnel should exercise safe working practice. Chemicals should be handled with suitable caution until their hazard(s) has been established.

b) Personnel should exercise health precautions to minimise risk to themselves and to ensure the integrity of the study.

c) Personnel known to have a health or medical condition that is likely to have an adverse effect on the study should be excluded from operations that may affect the study.

4.5.2 Quality Assurance Programme

General

a) The test facility should have a documented quality assurance programme to ensure that studies performed are in compliance with these Principles of Good Laboratory Practice.

b) The quality assurance programme should be carried out by an individual or by individuals designated by and directly responsible to management and who are familiar with the test procedures.

c) This individual(s) should not be involved in the conduct of the study being assured.

d) This individual(s) should report any findings in writing directly to management and to the Study Director.

Responsibilities of the Quality Assurance Personnel

The responsibilities of the quality assurance personnel should include, but not be limited to, the following functions:

a) ascertain that the study plan and Standard Operating Procedures are available to personnel conducting the study;

b) ensure that the study plan and Standard Operating Procedures are followed by periodic inspections of the test facility and/or by auditing the study in progress. Records of such procedures should be retained;

c) promptly report to management and the Study Director unauthorised deviations from the study plan and from Standard Operating Procedures;

d) review the final reports to confirm that the methods, procedures, and observations are accurately described, and that the reported results accurately reflect the raw data of the study;

e) prepare and sign a statement, to be included with the final report, which specifies the dates inspections were made and the dates any findings were reported to management and to the Study Director.

4.5.3 Facilities

General

a) The test facility should be of suitable size, construction and location to meet the requirements of the study and minimise disturbances that would interfere with the validity of the study.

b) The design of the test facility should provide an adequate degree of separation of the different activities to assure the proper conduct of each study.

Test System Facilitities

a) The test facility should have a sufficient number of rooms or areas to assure the isolation of test systems and the isolation of individual projects, involving substances known or suspected of being biohazardous.

b) Suitable facilities should be available for the diagnosis, treatment and control of diseases, in order to ensure that there is no unacceptable degree of deterioration of test systems.

c) There should be storage areas as needed for supplies and equipment. Storage areas should be separated from areas housing the test systems and should be adequately protected against infestation and contamination. Refrigeration should be provided for perishable commodities.

Facilities for Handling Test and Reference Substances

a) To prevent contamination or mix-ups, there should be separate areas for receipt and storage of the test and reference substances, and mixing of the test substances with a vehicle.

b) Storage areas for the test substances should be separate from areas housing the test systems and should be adequate to preserve identity, concentration, purity, and stability, and ensure safe storage for hazardous substances.

Archive Facilities

a) Space should be provided for archives for the storage and retrieval of raw data, reports, samples, and specimens.

Waste Disposal

a) Handling and disposal of wastes should be carried out in such a way as not to jeopardise the integrity of studies in progress.

b) The handling and disposal of wastes generated during the performance of a study should be carried out in a manner which is consistent with pertinent regulatory requirements. This would include provision for appropriate collection, storage, and disposal facilities, decontamination and transportation procedures, and the maintenance of records related to the preceding activities.

4.5.4 Apparatus, material, and reagents

Apparatus

a) Apparatus used for the generation of data, and for controlling environmental factors relevant to the study should be suitably located and of appropriate design and adequate capacity.

b) Apparatus used in a study should be periodically inspected, cleaned, maintained, and calibrated according to Standard Operating Procedures. Records of procedures should be maintained.

Material

a) Apparatus and materials in studies should not interfere with the test systems.

Reagents

    a) Reagents should be labelled, as appropriate, to indicate source, identity, concentration, and stability information and should include the preparation date, earliest expiration date and specific storage instructions.

4.5.5 Test Systems

Physical/Chemical

    a) Apparatus used for the generation of physical/chemical data should be suitably located and of appropriate design and adequate capacity.

    b) Reference substances should be used to assist in ensuring the integrity of the physical/chemical test systems.

Biological

    a) Proper conditions should be established and maintained for the housing, handling and care of animals, plants, microbial as well as other cellular and sub-cellular systems, in order to ensure the quality of the data.

    b) In addition, conditions should comply with appropriate national regulatory requirements for the import, collection, care and use of animals, plants, microbial as well as other cellular and sub-cellular systems.

    c) Newly received animal and plant test systems should be isolated until their health status has been evaluated. If any unusual mortality or morbidity occurs, this lot should not be used in studies and, when appropriate, humanely destroyed.

    d) Records of source, date of arrival, and arrival condition should be maintained.

    e) Animal, plant, microbial, and cellular test systems should be acclimatised to the test environment for an adequate period before a study is initiated.

    f) All information needed to properly identify the test systems should appear on their housing or containers.

    g) The diagnosis and treatment of any disease before or during a study should be recorded.

4.5.6 Test and Reference Substances

Receipt, Handling, Sampling and Storage

    a) Records including substance characterisation, date of receipt, quantities received and used in studies should be maintained.

    b) Handling, sampling and storage procedures should be identified in order that the homogeneity and stability is assured to the degree possible and contamination or mix up are precluded.

    c) Storage container(s) should carry identification information, earliest expiration date, and specific storage instructions.

Characterisation

    a) Each test and reference substance should be appropriately identified (e.g. code, chemical abstract number (CAS), name).

b) For each study, the identity, including batch number, purity, composition, concentration, or other characterisations to appropriately define each batch of the test or reference substances should be known.

c) The stability of test and reference substances under conditions of storage should be known for all studies.

d) If the test substance is administered in a vehicle Standard Operating Procedures should be established for testing the homogeneity and stability of the test substance in that vehicle.

e) A sample for analytical purposes from each batch of test substance should be retained for studies in which the test substance is tested longer than four weeks.

4.5.7 Standard Operating Procedures

General

a) A test facility should have written Standard Operating Procedures approved by management that are intended to ensure the quality and integrity of the data generated in the course of the study.

b) Each separate laboratory unit should have immediately available Standard Operating Procedures relevant to the activities being performed therein. Published text books, articles and manuals may be used as supplements to these Standard Operating Procedures.

Application

Standard Operating Procedures should be available for, but not be limited to, the following categories of laboratory activities. The details given under each heading are to be considered as illustrative examples.

i) Test and Reference Substances

Receipt, identification, labelling, handling, sampling and storage.

ii) Apparatus and Reagents

Use, maintenance, cleaning, calibration of measuring apparatus and environmental control equipment; preparation of reagents.

iii) Record keeping, reporting, storage, and retrieval

Coding of studies, data collection, preparation of reports, indexing systems, handling of data, including the use of computerised data systems.

iv) Test system (where appropriate)

Room preparation and environmental room conditions for the test system.

Procedures for receipt, transfer, proper placement, characterisation, identification and care of test system.

Test system preparation, observations and examinations, before, during and at termination of the study.

Handling of test system individuals found moribund or dead during the study.

Collection, identification and handling of specimens including necropsy and histopathology.

v) Quality Assurance Procedures

Operation of quality assurance personnel in performing and reporting study audits, inspections, and final study report reviews.

vi) Health and Safety Precautions

As required by national and/or international legislation or guidelines.

4.5.8 Performance of the study

Study Plan

a) For each study, a plan should exist in a written form prior to initiation of the study.

b) The study plan should be retained as "raw data".

c) All changes, modifications, or revisions of the study plan, as agreed to by the Study Director, including justification(s), should be documented, signed and dated by the Study Director and maintained with the study plan.

Content of the Study Plan

The study plan should contain, but not be limited to, the following information:

a) a descriptive title;

b) a statement which reveals the nature and purpose of the study;

c) identification of the test substance by code or name (IUPAC; CAS number, etc);

d) the reference substance to be used.

Information Concerning the Sponsor and the Test Facility

a) Name and address of the Sponsor.

b) Name and address of the Test Facility.

c) Name and address of the Study Director.

Dates

a) The date of agreement to the study plan by signature of the Study Director and, when appropriate, of the sponsor and/or the test facility management.

b) The proposed starting and completion dates.

Test Methods

a) Reference to OECD Test Guideline or other test guideline to be used.

Issues (where applicable)

a) The justification for selection of the test system.

b) Characterisation of the test system, such as the species, strain, sub-strain, source of supply, number, body weight range, sex, age, and other pertinent information.

c) The method of administration and the reason for its choice.

d) The dose levels and/or concentration(s), frequency, duration of administration.

e) Detailed information on the experimental design, including a description of the chronological procedure of the study, all methods, materials and conditions, type and frequency of analysis, measurements, observations and examinations to be performed.

Records

a) A list of records to be retained.

Conduct of the Study

A unique identification should be given to each study. All items concerning this study should carry this identification.

a) The study should be conducted in accordance with the study plan.

b) All data generated during the conduct of the study should be recorded directly, promptly, accurately, and legibly by the individual entering the data. These entries should be signed or initialled and dated.

c) Any change in the raw data should be made so as not to obscure the previous entry, and should indicate the reason, if necessary, for change and should be identified by date and signed by the individual making the change.

d) Data generated as a direct computer input should be identified at the time of data input by the individual(s) responsible for direct data entries. Corrections should be entered separately accompanied by the reason for change, with the date and the identity of the individual making the change.

4.5.9 Reporting of Study Results

General

a) A final report should be prepared for the study.

b) The use of Standard International Units is recommended.

c) The final report should be signed and dated by the Study Director.

d) If reports of principal scientists from co-operating disciplines are included in the final report, they should sign and date them.

e) Corrections and additions to a final report should be in the form of an amendment. The amendment should clearly specify the reason for the corrections or additions and should be signed and dated by the Study Director and by the principal scientist from each discipline involved.

Content of the Final Report

The final report should include, but not be limited to, the following information:

Identification of the Study, the Test and Reference Substance

a) A descriptive title.

b) Identification of the test substance by code or name (IUPAC; CAS number, etc).

c) Identification of the reference substance by chemical name.

d) Characterisation of the test substance including purity, stability and homogeneity.

Information Concerning the Test Facility

a) Name and address.

b) Name of the Study Director.

c) Name of other principal personnel having contributed reports to the final report.

Dates

a) Dates on which the study was initiated and completed.

Statement

a) A Quality Assurance statement certifying the dates inspections were made and the dates any findings were reported to management and to the Study Director.

Description of Materials and Test Methods

a) Description of methods and materials used.

b) Reference to ORCD Test Guidelines or other test guidelines.

Results

a) A summary of results.

b) All information and data required in the study plan.

c) A presentation of the results, including calculations and statistical methods.

d) An evaluation and discussion of the results and, where appropriate, conclusions.

Storage

a) The location where all samples, specimens, raw data, and the final report are to be stored.

4.5.10 Storage and Retention of Records and Material

Storage and Retrieval

a) Archives should be designed and equipped for the accommodation and the secure storage of:

the study plans;

the raw data;

the final reports;

the reports of laboratory inspections and study audits performed according to the Quality Assurance Programme;

samples and specimens.

b) Material retained in the archives should be indexed so as to facilitate orderly storage and rapid retrieval.

Only personnel authorised by management should have access to the archives. Movement of material in and out of the archives should be properly recorded.

Retention

a) The following should be retained for the period specified by the appropriate authorities:

the study plan, raw data, samples, specimens, and the final report of each study;

records of all inspections and audits performed by the Quality Assurance Programme;

Summary of qualifications, training, experience and job descriptions of personnel;

Records and reports of the maintenance and calibration of equipment;

The historical file of Standard Operating Procedures.

b) Samples and specimens should be retained only as long as the quality of the preparation permits evaluation.

c) If a test facility or an archive contracting facility goes out of business and has no legal successor, the archive should be transferred to the archives of the sponsor(s) or the study(s).

4.6 Concluding remarks

As you will see, the principles of Good Laboratory Practice are extensive. The purpose of Good Laboratory Practice is of course to promote the development of quality test data. Comparable quality of test data forms the basis for the mutual acceptance of test data among countries.

Because individual countries can confidently rely on test data developed in other countries, duplicative testing can be avoided, thereby introducing economies in test costs and time. The application of these Principles has helped to avoid the creation of technical barriers to trade, and further improves the protection of human health and the environment.

Guidelines for microbial and animal cell cultivation

Appendix 5.1

Guidelines for microbial and animal cell cultivation

5.1 Introduction

In this chapter, we discuss the safety issues associated with the use of micro-organisms. This reflects both the importance of micro-organisms in biotechnology and their importance as potential mediators of diseases of mankind and other biological entities. In the following chapter, we will discuss the safety issues concerned with genetically manipulating organisms and the assessment of the risks involved in constructing organisms displaying changed characteristics.

We will not discuss the mechanisms by which humans may be harmed by pathogenic micro-organism but confine ourselves to aspects of safety regulations and law including the categorisation of pathogens according to hazards and categories of containment that are required to prevent the escape of organisms.

We begin by considering the relevant EC-Directives and then extend the discussion to include some specific examples of differences in national practices. We will also examine the containment measures that need to be taken when working with particular hazard groups of organisms. Towards the end of the chapter we will include a discussion of the special issues associated with the cultivation of animal cells.

5.2 The Legislation - the Council Directive on the Protection of Workers from Risks Related to Exposure to Biological Agents at Work

Key Directive   The most important EC legislation which deals with the protection of workers and other individuals from risks associated with the cultivation of potentially dangerous micro-organisms is contained within a Directive made in 1990 (90/679/EEC) entitled "Council Directive on the protection of workers from risks related to exposure to biological agents at work (seventh individual directive within the meaning of article 16 (1) of Directive 89/391/EEC)".

This Directive has 20 Articles grouped into three main sections (section I, General Provisions; section II, Employers' Obligations; section III, Miscellaneous Provisions). The Directive is a development of earlier EC principals, in particular Article 118a of the Treaty establishing the European Economic Community which "provides that the Council shall adopt, by means of Directives, minimum requirements in order to encourage improvements in the working environment, so as to guarantee better protection of the health and safety of workers". This includes, in this particular legislation, the development of protective and preventive measures for workers exposed to dangerous biological agents and an onus on employers to develop good practice to gain knowledge of the risks involved in exposure to biological agents, for example, through the keeping of records and through keeping abreast of new developments in technology.

In making this Directive, the Commission consulted the EC Advisory Committee on Safety, Hygiene and Health Protection at Work. The main overall aim of the Directive is to protect workers against risks to their health and safety, by preventing such risks, arising or likely to arise from exposure to biological agents at work. The Directive lays down particular minimum provisions in this area and applies "without prejudice" to the provisions of previous Council Directives 20/220/EEC (on the deliberate release into the environment of genetically modified organisms).

5.2.1 Definitions

Certain definitions are given in Article 2:

a) 'biological agents' shall mean micro-organisms, including those which have been genetically modified, cell cultures and human endoparasites, which may be able to provoke any infection, allergy or toxicity;

b) 'micro-organism' shall mean a microbiological entity, cellular or non-cellular, capable of replication or of transferring genetic material;

c) 'cell culture' shall mean the *in-vitro* growth of cells derived from multicellular organisms;

d) 'biological agents' shall be classified into four risk groups, according to their level of risk of infection:

Group 1 biological agent is one that is unlikely to cause human disease;

Group 2 biological agent is one that can cause human disease and might be a hazard to workers, but is unlikely to spread to the community; there is usually effective prophylaxis or treatment available;

Group 3 biological agent is one that can cause severe human disease and present a serious hazard to workers; it may present a risk of spreading to the community, but there is usually effective prophylaxis or treatment available;

Group 4 biological agent is one that causes severe human disease and is a serious hazard to workers; it may present a high risk of spreading to the community; there is usually no effective prophylaxis or treatment available.

We will give examples of organisms in each of these groups at the end of the chapter.

5.2.2 Determination and assessment of risks (Article 3)

This states that:

a) in the case of any activity likely to involve a risk of exposure to biological agents, the nature, degree and duration of workers' exposure must be determined in order to make it possible to assess any risk to the workers' health or safety and to lay down the measures to be taken;

b) in the case of activities involving exposure to several groups of biological agents, the risk shall be assessed on the basis of the danger presented by all hazardous biological agents present;

c) the assessment must be renewed regularly and in any event when any change occurs in the conditions which may affect workers' exposure to biological agents;

d) the employer must supply the competent authorities, at their request, with the information used for making the assessment.

The assessment earlier referred to has to be conducted on the basis of all available information including:

i) classification of biological agents which are or may be a hazard to human health (as referred to in Article 18);

ii) recommendations from a competent authority which indicate that the biological agent should be controlled in order to protect workers' health when workers are or may be exposed to such a biological agent as a result of their work;

iii) information on diseases which may be contracted as a result of the work of the workers;

iv) potential allergenic or toxigenic effects as a result of the work of the workers;

v) knowledge of a disease from which a worker is found to be suffering and which has a direct connection with his work.

5.2.3 Employers obligations

The second section of the EC legislation concerns employers' obligations. Thus, the employer shall avoid the use of a harmful biological agent if the nature of the activity so permits, by replacing it with a biological agent which, under its conditions of use, is not dangerous or is less dangerous to workers' health. With regard to the reduction of risks, if the results of the assessment (referred to in Article 3) reveal a risk to health or safety, then workers' exposure must be prevented or (if not technically practicable) the risk of exposure must be reduced to as low a level as necessary in order to protect adequately the health and safety of the workers concerned. The following measures are suggested:

a) keeping as low as   possible the number of workers exposed or likely to be exposed;

b) design of work processes and engineering control measures so as to avoid or minimise the release of biological agents into the place of work;

c) collective protection measures and/or, where exposure cannot be avoided by other means, individual protection measures;

d) hygiene measures compatible with the aim of the prevention or reduction of the accidental transfer or release of a biological agent from the workplace;

e) use of the biohazard sign (depicted in the annex) and other relevant warning signs;

f) drawing up plans to deal with accidents involving biological agents;

g) testing, where it is necessary and technically possible, for the presence, outside the primary physical confinement, of biological agents used at work;

h) means for safe collection, storage and disposal of waste by workers, including the use of secure and identifiable containers, after suitable treatment where appropriate;

i) arrangements for the safe handling and transport of biological agents within the workplace.

5.2.4 Interaction with the competent authority

Article 7 of the legislation relates to the information that must be given to the competent authority should the results of the assessment (referred to in Article 3) reveal a risk to workers' health or safety. Thus, "employers shall, when requested, make available to the competent authority appropriate information on:- the results of the assessment; the activities in which workers have been exposed or may have been exposed to biological agents; the number of workers exposed; the name and capabilities of the person responsible for safety and health at work; the protective and preventive measures taken, including working procedures and methods; an emergency plan for the protection of workers from exposure to a group 3 or a group 4 biological agent which might result from a loss of physical containment". Moreover, employers must inform the competent authority of any accident or incident which may have resulted in the release of a biological agent and which could cause severe human infection and/or illness. A list of exposed workers (referred to at a later point in Article 11) and the medical record (referred to in Article 14) must be made available to the competent authority in accordance with national laws and/or practice in cases where the undertaking ceases activity .

5.2.5 Hygiene and protection

Article 8 is concerned with hygiene and protection of individuals and states that:

1) Employers shall be obliged, in the case of all activities for which there is a risk to the health or safety of workers due to work with biological agents, to take appropriate measures to ensure that:

   a) workers do not eat or drink in working areas where there is a risk of contamination by biological agents;

   b) workers are provided with appropriate protective clothing or other appropriate special clothing;

   c) workers are provided with appropriate and adequate washing and toilet facilities, which may include eye washes and/or skin antispectics;

   d) any necessary protective equipment is: properly stored in a well-defined place; checked and cleaned if possible before, and in any case after, each use; is repaired, when defective, or is replaced before further use;

   e) procedures are specified for taking, handling and processing samples of human or animal origin.

2) a) Working clothes and protective equipment, including protective clothing referred to previously, must be removed on leaving the working area and, before taking the measures referred to in (b), kept separately from other clothing.

   b) The employer must ensure that such clothing and protective equipment is decontaminated and cleaned or, if necessary, destroyed.

3) Workers may not be charged for the cost of the measures referred to in paragraphs 1 and 2 above.

5.2.6 Training requirements

Article 9 covers information and training of workers and outlines the responsibilities of the employer to ensure that workers receive sufficient and appropriate training, on the basis of all available information, concerning:

    a) potential risks to health;

    b) precautions to be taken to prevent exposure;

    c) hygiene requirements;

    d) wearing and use of protective equipment and clothing;

    e) steps to be taken by workers in the case of incidents and how to prevent incidents.

With regard to training it states that this should be given at the beginning of work involving contact with biological agents, adapted to take account of new or changed risks and repeated periodically if necessary.

5.2.7 Adoption of written procedures

Article 10 refers to worker information which is necessary in particular cases. Thus, it is the duty of employers to provide written instructions (eg. to display notices) of the procedure to be followed in the case of a serious accident or incident involving the handling of a biological agent or when handling a group 4 biological agent. Workers have a duty to immediately report any accident or incident involving the handling of a biological agent to the person in charge or to the person responsible for safety and health at work and employers must inform workers (and/or any workers' representatives) of any accident or incident which may have resulted in the release of a biological agent and which could cause severe human infection and/or illness. In addition, information must be given as quickly as possible when a serious accident or incident occurs, of the causes thereof, and of the measures taken or to be taken to rectify the situation. Article 10 also states that "Each worker shall have access to the information on the list referred to in Article 11 which relates to him personally" and that "Workers and/or any workers' representatives in the undertaking or establishment shall have access to anonymous collective information". Employers must also provide workers and/or their representatives, at their request, with the information provided for in Article 7(1) which concerns the results of assessment.

5.2.8 Records of exposed workers

Article 11 sets out the rules to be followed regarding the listing of exposed workers. Thus:

    1)   Employers shall keep a list of workers exposed to group 3 and/or group 4 biological agents, indicating the type of work done and, whenever possible, the biological agent to which they have been exposed, as well as records of exposures, accidents and incidents, as appropriate.

    2)   The list referred to in paragraph 1 shall be kept for at least 10 years following the end of exposure, in accordance with national laws and/or practice. In the case of those exposures which may result in infections:- with biological agents known to be capable of establishing persistent or latent infections,- that, in the light of

present knowledge, are undiagnosable until illness develops many years later,- that have particularly long incubation periods before illness develops,- that result in illnesses which recrudesce at times over a long period despite treatment, or- that may have serious long-term sequelae, the list shall be kept for an appropriately longer time up to 40 years following the last known exposure.

3) The doctor (referred to in Article 14) and/or the competent authority for health and safety at work, and any other person responsible for health and safety at work, shall have access to the list referred to in paragraph 1.

5.2.9 Worker participation and notification made to competant authorities

Article 12 covers aspects of consultation and participation of workers and/or their representatives in connection with matters covered by the Directive, and Article 13 refers to the requirements of notification to the competent authority. Thus:

1) Prior notification shall be made to the competent authority of the use for the first time of: group 2 biological agents,- group 3 biological agents,- group 4 biological agents. The notification shall be made at least 30 days before the commencement of the work. Subject to paragraph 2, prior notification shall also be made of the use for the first time of each subsequent group 4 biological agent and of any subsequent new group 3 biological agent where the employer himself provisionally classifies that biological agent.

2) Laboratories providing a diagnostic service in relation to group 4 biological agents shall be required only to make an initial notification of their intention.

3) Renotification must take place in any case where there are substantial changes of importance to safety or health at work to processes and/or procedures which render the notification out of date.

4) The notification referred to in this Article shall include: a) the name and address of the undertaking and/or establishment; b) the name and capabilities of the person responsible for safety and health at work; (c) the results of the assessment referred to in Article 3; d) the species of the biological agent; e) the protection and preventive measures that are envisaged.

5.2.10 Health surveillance

The final section of this particular legislation deals with "Miscellaneous Provisions". Article 14 concerns "Health surveillance" and makes it a duty of Member States to "establish, in accordance with national laws and practice, arrangements for carrying out relevant health surveillance of workers for whom the results of the assessment referred to in Article 3 reveal a risk to health or safety". The "arrangements" referred to must be such that workers can undergo, if appropriate, relevant health surveillance prior to exposure and at regular intervals thereafter. Moreover, the arrangements shall be such that it is directly possible to implement individual and occupational hygiene measures. Article 14 also refers to those workers for whom special protection measures may be required and states that "When necessary, effective vaccines should be made available for those workers who are not already immune to the biological agent to which they are exposed or are likely to be exposed". If a worker is found to be suffering from an infection and/or illness which is suspected to be the result of exposure, the doctor or authority responsible for health surveillance of workers must offer surveillance to other workers who have been similarly exposed. In this event, a reassessment of the risk of exposure must be carried out in accordance with Article 3.

Article 14 also covers the duties of the medical doctor or authority responsible for health surveillance. With regard to the keeping of medical surveillance records it states that these "shall be kept for at least 10 years following the end of exposure, in accordance with national laws and practice". In special cases an individuals medical record must be kept for an appropriately longer time up to 40 years following the last known exposure. The doctor or authority responsible for health surveillance must also propose any protective or preventive measures to be taken in respect of any individual worker and give information and advice to workers regarding any health surveillance which they may undergo following exposure. The results of health surveillance which concern individuals must be made available to them upon request and both the workers concerned or the employer may request a review of the results of the health surveillance. It is also a duty of the doctor or authority responsible for health surveillance to report all cases of diseases or death to the competent authority in accordance with national laws and/or practice.

5.2.11 Measures to respond to known pathogens in biological specimens

Article 15 concerns health and veterinary care facilities other than diagnostic laboratories and states that:

1)   For the purpose of the assessment referred to in Article 3, particular attention should be paid to: a) uncertainties about the presence of biological agents in human patients or animals and the materials and specimens taken from them; b) the hazard represented by biological agents known or suspected to be present in human patients or animals and materials and specimens taken from them; c) the risks posed by the nature of the work.

2)   Appropriate measures shall be taken in health and veterinary care facilities in order to protect the health and safety of the workers concerned. The measures to be taken shall include in particular:

    a) specifying appropriate decontamination and disinfection procedures, and

    b) implementing procedures enabling contaminated waste to be handled and disposed of without risk.

3)   In isolation facilities where there are human patients or animals who are, or who are suspected of being, infected with group 3 or group 4 biological agents, containment measures shall be selected from those in Annex V column A, in order to minimize the risk of infection.

5.2.12 Measures relating to deliberately infected animals

Article 16 concerns special measures for industrial processes, laboratories and animal rooms where laboratory animals which have been deliberately infected with group 2, 3 or 4 biological agents or which are or are suspected to be carriers of such agents are kept.

containment measures to restrict contamination

It states that the laboratories "shall determine the containment measures in accordance with Annex V, in order to minimize the risk of infection". Moreover, "measures shall be determined in accordance with Annex V, after fixing the physical containment level required for the biological agents according to the degree of risk". Thus, activities involving the handling of a biological agent must only be carried out in working areas corresponding to at least containment level 2, for a group 2 biological agent, to at least containment level 3, for a group 3 biological agent and only in containment level 4, for a group 4 biological agent. Those laboratories handling materials in respect of which there exist uncertainties about the presence of biological agents which may cause human disease but which do not have as their aim working with biological agents as such (i.e. cultivating or concentrating them) must adopt containment level 2 at least and containment levels 3 or 4 must be used, when appropriate, where it is known or it is suspected that they are necessary, except where guidelines provided by the competent national authorities show that, in certain cases, a lower containment level is appropriate.

5.2.13 Containment principles in industrial processes

Article 16 also stipulates that industrial processes using group 2, 3 or 4 biological agents must adhere to the containment principles already set out in this particular legislation and that following assessment of risk, appropriate measures must be applied as seems fit. If it is not possible to carry out a conclusive assessment of a biological agent even though it appears that the envisaged use may involve a serious health risk, then activities may be restricted to workplaces where the containment level corresponds at least to level 3.

Article 17 states that "The Commission shall have access to the use made by the competent national authorities of the information referred to in Article 14 (9)".

5.2.14 Classification of biological agents

Article 18 makes it clear that the Council shall adopt a preliminary list of group 2, group 3 and group 4 biological agents (listed in Annex III) and that the Community classification will be based on the definitions outlined in Article 2 d) points 2 to 4 (groups 2 to 4). It also requests that Member States shall classify biological agents that are or may be a hazard to human health on the same basis as defined in Article 2 d) points 2 to 4 (groups 2 to 4). If the biological agent to be assessed cannot be classified clearly in one of the groups defined in Article 2 d), it must be classified in the highest risk group among the alternatives.

Article 19 concerns the annexes and explains that purely technical adjustments to the annexes may be required in the light of technical progress, changes in international regulations or specifications and new findings in the field of biological agents.

Article 20 (final provisions) states when the Member States must bring the Directive into force (ie before December 1993).

5.3 Comparison of classification of micro-organisms into risk (hazard) groups

In Section 5.2.1 , we described the division of biological agents into four groups. These groups more or less follow the four risk classes defined by the European Federation of Biotechnology (EFB) in 1989. The EFB groups are:

- harmless micro-organisms;

- low-risk micro-organisms;

- medium-risk micro-organisms;

- high-risk micro-organisms.

A description of these categories is provided in Table 5.1.

The four EFB risk classes are reflected in many national-based groupings. Here we will give just two examples. The UK Advisory Committee on Dangerous Pathogens (ACDP) is an agency of the UK Health and Safety Executive. They define four hazard groups (see Table 5.2) which have great similarity to those defined by EFB. Similarly in The Netherlands, four groups are defined (Pathogen Groups 1,2,3,4 - often abbreviated to PG1,2,3,4).

Harmless micro-organisms have an extended history of safe use in human consumption (eg beer, yoghurt, cheese) and on a large scale in the biotechnology industry (antibiotics, enzymes, amino acids) as well as in agricultural applications (*Bacillus thuringiensis*).
EFB Class 1 is defined as:
'Those micro-organisms that never have been identified as causative agents of disease in Man and that offer no threat to the environment. They are not listed in higher classes'.

Low-risk micro-organisms do not easily survive outside the host, are unlikely to spread and do not cause serious diseases.
EFB Class 2 is defined as:
'Those micro-organisms that may cause disease in Man and which might , therefore, offer a hazard to laboratory workers.
They are unlikely to spread in the environment.
Prophylactics are available and treatment is effective.'

Medium- and **high-risk** micro-organisms are severe pathogens and they are rarely used for large-scale operations. On a small scale, they are used for the production of vaccines or diagnostics.
EFB Class 3 is defined as:
'Those micro-organisms that offer a severe threat to the health of laboratory workers but a comparatively small risk to the population at large. Prophylactics are available and treatment is effective.'
EFB Class 4 is defined as:
'Those micro-organisms that cause severe illness in Man and offer a serious hazard to laboratory workers and to the population at large'.

Table 5.1 EFB risk classes of naturally-occurring micro-organisms.

Table 5.2 shows ACDP hazard groups. Compare these groupings with those given in Table 5.1.

Hazard Group 1

An organism that is most unlikely to cause human disease.

Hazard Group 2

An organism which may cause human disease and which might be a hazard to laboratory workers but is unlikely to spread to the community.

Laboratory exposure rarely produces infection and effective prophylaxis or effective treatment is usually available.

Hazard Group 3

An organism that may cause serious human disease and presents a serious hazard to laboratory workers. It may present a risk to the community. Usually effective prophylaxis or treatment is not available.

Table 5.2 ACDP hazard groups.

The lists of organisms in the risk classes (hazard groups) are very long and cover: bacteria, chlamydias, rickettsias, mycoplasmas, fungi, parasites and viruses. Certain strains of organisms which are listed may justify different levels of containment than the hazard ratings suggest. This applies to mutants of increased or decreased virulence, attenuated vaccine viruses or bacteria and antibiotic resistant strains (it also applies to genetically manipulated bacteria. In cases where many organisms have been grouped together certain members (species or strains) may justify the use of different levels of containment than is necessary for the group as a whole.

different
restrictions
may apply in
different states

We have provided examples of specific organisms from some of the Hazard Groups in Appendix 5.1 at the end of the chapter. This list is based on the recommendations of the ACDP in the UK. There is, however, more-or-less international agreement on the hazards (risks) associated with the named examples. You should, of course, become acquainted with the specific details of the recommendations in your region. Particular states impose specific conditions on some pathogens. For example, deliberate cultivation of the *Variola* (smallpox) virus is banned totally in some countries. The reader should not assume that the absence of a particular species from our list means that the organism is free from risk. A key aspect to working with micro-organisms is the need to be aware, not only of the biological hazards but also of the regulatory obligations that must be fulfilled.

plant pathogens

It is not only human pathogens that are covered by such regulations. Similar restrictions apply to work with plant pathogens. These too are categorised, although the risk of plant pathogens depends more on local situations, for example is the plant pathogen endemic in the area or is it exotic? High-risk plant pathogens are quarantine organisms. In the UK for example, it is the responsibility of the user to determine whether or not the organism they wish to use is a plant pathogen or pest, whether or not it is indigenous, and to apply for a licence to use the organism if appropriate. Where possible the user will be helped to select a lower risk organism.

The EFB has published for plant pathogens, a risk classification (Küenz *et al* 1987, Appl Microbial Biotechnol 27,105).

There are three classes: low-, medium- and high-risk. For low- and medium-risk pathogens, the Plant Protection Authority should always be consulted. High-risk plant pathogens are quarantine organisms; they should not be used (see Table 5.3).

Ep1	low-risk if not endemic
Ep2	medium-risk for crops
Ep3	high-risk quarantine organisms

Table 5.3 EFB classes of plant pathogens.

Similarly, the use and importation of any pathogen that may cause disease in agricultural animals, birds, fish and bees is strictly controlled.

Regulatory aspects of the importation/use of these pathogens vary from county to country and are the subject of continuous updating.

5.4 Laboratory practice

In the remainder of this chapter, we will concentrate on the requirements for work at various levels of containment.

At the entrance to laboratories carrying out work with biologically hazardous material the sign shown in Figure 5.1 should be shown.

Figure 5.1 International Biohazard sign.

5.5 Classes of containment

In the previous sections, we have explained that micro-organisms can be divided into four risk classes (hazard groups). The extent of containment that needs to be maintained reflects the risks associated with each of these groups. As a general principle, the chance of escape should be **minimised** for low-risk micro-organisms and should be **prevented** for medium and high-risk micro-organisms. These safety precautions can be divided into three aspects:

- procedures;

- primary containment (equipment);

- secondary containment (facilities).

harmonisation of national regulation

Again there are national differences in the details of the recommendations and regulations which are applied to these issues although they have many features in common. International bodies such as European Federation of Biotechnology (EFB) and the Organisation for Economic Co-operation and Development (OECD) are doing much to harmonise the different national risk classifications and containment systems. Similarly the EC-Directives 90/679/EEC, described in Section 5.2, will do much to harmonise national approches. Here we will take a generally accepted position. It is, however, important that readers are familiar with their own national situation. For example, the current Netherlands Recommendations for Safe Microbiological Work are published in Aanbevelingen voor Veilig Microbiologisch Werk 2nd ed. Editors: Nederlandse Vereniging voor Microbiologie, RIVM Bilthoven. In the UK these are published by HMSO (Her Majesty's Stationery Office) in a publication entitled, 'Advisory Committee on Dangerous Pathogens - Categorisation of Pathogens According to Hazard and Categories of Containment 1990'.

5.5.1 Harmless micro-organisms (EFBI, Hazard Group 1, PGI)

For the use of harmless micro-organisms, there exist only limited containment requirements.

BMT

Such organisms can, for example, be used in industrial processes such as food manufacture. In such circumstances, general, good microbiological practices are usually sufficient. These are required to protect cultures from contamination and to provide quality assurance. Similarly in the laboratory, basic microbiological techniques (BMT) are sufficient to ensure the absence of contaminants (use of sterile culture media, aseptic inoculation). We have provided a model set of laboratory rules for the containment of micro-organisms in this category in Table 5.4. We will call this Containment Level 1 (consistent with ACDP nomenclature).

GILSP

When handling this group of organisms on an industrial scale, for example in the manufacture of such foods as yoghurt, cheese, mycoprotein and food enzymes, analogous standards are applied. A set of measures called Good Industrial Large-Scale Practices (GILSP) have been devised by the OECD.

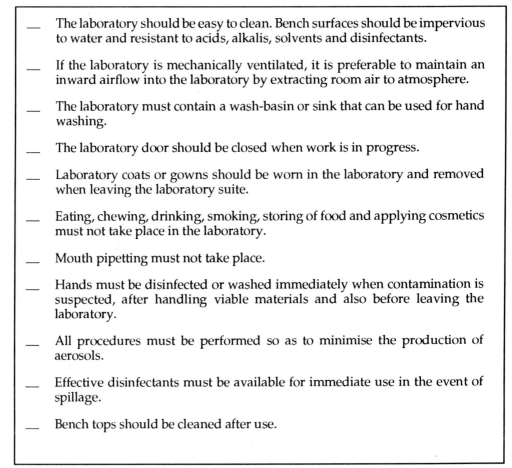

> — The laboratory should be easy to clean. Bench surfaces should be impervious to water and resistant to acids, alkalis, solvents and disinfectants.
>
> — If the laboratory is mechanically ventilated, it is preferable to maintain an inward airflow into the laboratory by extracting room air to atmosphere.
>
> — The laboratory must contain a wash-basin or sink that can be used for hand washing.
>
> — The laboratory door should be closed when work is in progress.
>
> — Laboratory coats or gowns should be worn in the laboratory and removed when leaving the laboratory suite.
>
> — Eating, chewing, drinking, smoking, storing of food and applying cosmetics must not take place in the laboratory.
>
> — Mouth pipetting must not take place.
>
> — Hands must be disinfected or washed immediately when contamination is suspected, after handling viable materials and also before leaving the laboratory.
>
> — All procedures must be performed so as to minimise the production of aerosols.
>
> — Effective disinfectants must be available for immediate use in the event of spillage.
>
> — Bench tops should be cleaned after use.

Table 5.4 Model laboratory rules for Containment Level 1. These rules are typical of those which are applied to handling micro-organisms classified as being harmless (eg EFB1, UK-Hazard Group 1, Netherlands-PG1). These rules are based on those recommended by the Advisory Committee on Dangerous Pathogens in the UK and are designed to minimise the **release** of micro-organisms during laboratory work.

5.5.2 Low-risk micro-organisms (EFB2, Hazard Group 2, PG2).

GMT

Good Microbiological Techniques (GMT) are essential in all work with low and higher risk micro-organisms in the laboratory. In contrast with BMT, which is mainly concerned with ensuring that cultures are transferred without contamination, GMT protects the laboratory worker and the environment against contamination and possible infection from cultured pathogens.

Many of the features shown in Table 5.4 are also operational at this level. There are, however, some important additions. For example, in the UK it is recommended that $24m^3$ of air space is available for each worker, wash basin taps must be of a type that can be operated without being touched by hand and bench tops must be disinfected after use. It is also demanded that for manipulations such as vigorous shaking, ultrasonic disruption and other techniques that create aerosols, a microbiological safety cabinet or equipment designed to contain the aerosol must be used. There should also be restricted access to the facility.

In the UK, this level of containment is called Containment Level 2. EFB categorisation regards this level as Containment Category 1 (CC1). The equivalent in The Netherlands is Fysisch Inperkings Nivo 1 (FIN1).

5.5.3 Medium-risk micro-organisms (EFB3, Hazard Group 3, PG3)

This hazard group is regarded as Containment Category 2 (CC2) by EFB, as FIN2 in The Netherlands and Containment Level 3 in the UK. The conditions imposed under these categorisations are, however, very similar. We have highlighted the major measures that have to be taken in addition to those taken at lower containment levels in Table 5.5.

— only authorised personnel are admitted to the facility

— if vaccines are available, personnel are vaccinated

— if air is extracted from the facility, HEPA filters are used

— effluents from the facility should be decontaminated or sterilised

— an autoclave should be within the facility

— all processes involving medium-risk micro-organisms must be carried out in hermetically-sealed equipment or biosafety cabinets

— protective suits, closing at the back, have to be worn by personnel

---

Table 5.5 Measures taken in using medium-risk micro-organisms in addition to those taken with lower risk organisms. Note that with this category, the objective of the containment procedures is to **prevent** the release of micro-organisms during laboratory work. HEPA = High Efficiency Particulate Air.

5.5.4 High-risk micro-organisms (EFB4, Hazard Group 4, PG4)

Clearly use of organisms in this category must be done in such a way as to prevent any release. This implies prevention of release at **any** stage of procedures. Thus equipment and facilities must be designed to prevent any escape (ie full primary and secondary containment devices must be used). Only experienced personnel are allowed to handle such organisms and they must receive extensive training. Their work must be supervised. In Table 5.6 we highlight some features of this level of containment.

—   visitors are not admitted

—   facility must be completely isolated

—   the rooms for complete change of clothes must include an air lock facility
     with compulsory shower

—   negative pressure must be maintained in the facility

—   air ducts must be protected by double HEPA filters

—   materials containing high-risk micro-organisms must be absolutely
     separated from workers

Table 5.6 Measures taken in using high-risk micro-organisms in addition to those taken with lower risk
organisms. Note that under EFB nomenclature this level of containment is described as Containment
Category 3 (CC3). In The Netherlands it is described as FIN3 whilst in the UK it is referred to as
Containment Level 4. The aim of these measures is to be absolutely certain that organisms are **prevented**
from escaping from the laboratory.

Table 5.7 is a summary of the laboratory containment requirements for harmless (EFB
Class 1), low-risk (EFB Class 2), medium-risk (EFB Class 3) and high-risk (EFB Class 4)
micro-organisms. Y represents yes, meaning that the requirement is needed, N is for not
required.

Containment requirements	EFB Class of micro-organism			
	1	2	3	4
laboratory site:isolation	N	N	partial	Y
laboratory: sealable for fumigation	N	N	Y	Y
airlock	N	N	optional	Y
airlock with shower	N	N	N	Y
wash basin	Y	Y	Y	Y
effluent treatment	N	N	optional	Y
autoclave: in the laboratory suite	N	Y	Y	Y
in double-ended laboratory	N	N	N	Y

Table 5.7 Some features of containment requirements with micro-organisms of different risk classes.

5.6 Training

A key feature of all the various recommendations in operation is that they call for
suitable training of laboratory personnel. This training obviously involves instruction
concerning the details of operations and in most instances includes the use of 'biosafety
manuals' which give detailed procedural descriptions of operations being undertaken
in the laboratory. The extent of this training naturally increases with higher risk

biosafety
manuals

organisms. There is also a general requirement that an appropriate standard of supervision of the work is maintained.

Before we leave this aspect, we re-emphasise the point that it is essential that all workers are aware of the risk category of the micro-organisms they are handling and are familiar with both the appropriate national regulations and the local rules governing their use.

5.7 Measuring the safety of working conditions in laboratories and plants

5.7.1 Air contamination

RCS and CABS The amount of airborne contamination can be determined using the Reuter Centrifugal Sampler (RCS) or the Casella Airborne Bacteria Sampler (CABS). The former samples 40 litres of air per minute, the latter has a capacity of 700 litres of air per minute. These instruments may be used to confirm that safe limits for air contamination are being met.

5.7.2 Surface contamination

Bacterial contamination of surfaces in the workplace may be assessed using contact plates or by taking swab samples.

contact method In the contact method, RODAC (Replicate Organism Detecting and Counting) plates are filled with 'plate count agar' so that there is a slightly convex surface rising above the rim. The plates are applied to the surface to be tested (for example work-table or hand) and then incubated. Representative colonies are subcultured for identification.

swab method In the swab method, swabs are made by wrapping and binding cotton wool around the ends of glass rods. Each swab is wetted, placed into a test tube and sterilised. A defined part of the surface to be investigated is rubbed firmly with a wetted swab. The swab is then shaken with nutrient broth. Part of the nutrient broth is mixed with molten, nutrient agar and allowed to gel. Incubation follows and again representative colonies are identified. Results obtained from both methods correlate quite well but neither measures 100% of the surface contamination.

aerosol generation Using both surface and air monitoring methods, it has been shown that some processes such as pouring cultures into flasks, blowing out pipettes during work, subculturing, and open centrifugation caused severe air and surface contaminations. These could be avoided by carrying out such operations in a biosafety cabinet, which retains all aerosol droplets.

Human errors in the techniques used for handling micro-organisms may be discovered if monitoring is carried out consistently. Apart from the aerosol-creating procedures discussed another type of contamination which is sometimes found is that produced by steaming product lines without filtering the exhaust steam output, or cleaning equipment before thorough disinfection.

filter efficiencies The type of filter chosen to retain aerosolised bacteria must meet with the appropriate standards in the country of use. Filters may be tested by challenging them with known levels of bacteria and assessing the penetration of bacteria.

$$\text{Penetration} = \frac{\text{number of bacteria passing through the filter}}{\text{challenge number of bacteria}}$$

Obviously for high-risk organisms which have low ID50 values it is very important that all appropriate precautions are taken.

5.8 Safety cabinets

Microbiological safety cabinets are designed to capture and retain airborne particles released in the course of certain manipulations and hence protect the laboratory worker from infections which may arise from inhaling them.

There are three kinds or classes of safety cabinet. Again as you might anticipate, there are some differences between the fine details of the specifications of different classes of safety cabinets. Nevertheless, the standards set are very similar in many countries. Each country imposes its own specifications. For example in the UK, these cabinets should conform to British Standard 5726. Below we give a general description of the various classes of cabinets that are available and indicate when they should be used. Note that these cabinets are variously described as Class 1, 2 and 3 cabinets or Class I, II or III cabinets.

5.8.1 Class I cabinet

A Class I cabinet is shown in Figure 5.2 Air is drawn in from the open front over the working area, it is then filtered to remove infectious particles.

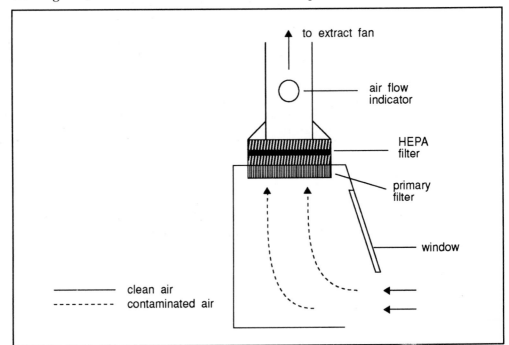

Figure 5.2 Class I microbiological safety cabinet, showing general design and airflow.

Typically, Class I cabinets are used with low-risk micro-organisms (EFB Class 2).

5.8.2 Class II cabinet

An example of a Class II cabinet is shown in Figure 5.3. Most of the air is recirculated through filters, some is dumped into the room and is replaced by air which is drawn through the open front. Class II cabinets protect the work within the cabinet from external contamination and provide some protection to operators.

In the UK, BS5726 has recently (from 1st January, 1993) been revised so that Class II cabinets dumping air into a laboratory must include dual in-line exhaust HEPA filters. This uprated standard, BS5726 (1992), is likely to be adopted across the EC.

This type of cabinet is used for medium-risk micro-organisms (EFB Class 3). Note that extensive precautions are taken to prevent the escape of organisms.

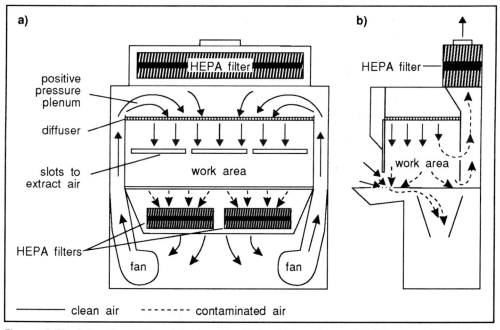

Figure 5.3 The Baker Class II Type 2 version of the National Cancer Institute's NCI-I cabinet. About 30% of the air is recirculated and 70% dumped after filtration. Air in the plenum has been filtered and is under positive pressure (courtesy of the Baker Company).

5.8.3 Class III cabinet

The cabinet is totally enclosed, the operator works via gloves attached to ports at the front. Air is filtered as it leaves the cabinet (see Figure 5.4). These cabinets are suitable for use in handling high-risk micro-organisms (EFB Class 4).

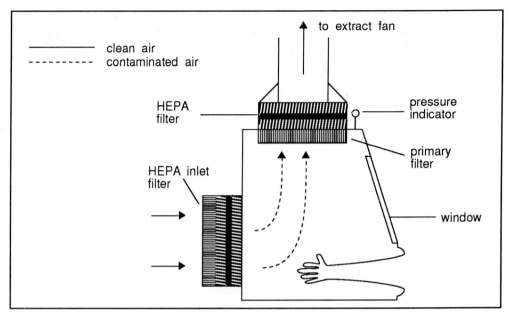

Figure 5.4 Class III microbiological safety cabinet, showing general design and airflow. Note that the micro-organisms and the worker are kept completely apart.

5.9 High efficiency particulate air (HEPA) filters

HEPA filters are made from glass fibre paper which is approximately 60 μm thick. Fibres vary from 0.4-1.4 μm in diameter. The filter is constructed by folding the sheet into a pleat within a box unit, see Figure 5.5.

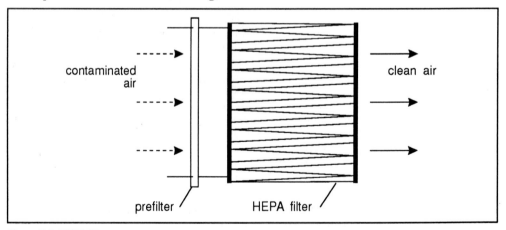

Figure 5.5 HEPA filter.

Biological safety cabinet HEPA filters are usually protected by coarser prefilters which remove dust and other particles down to about 5μm. These coarse filters are cheaper than HEPA filters and act to prolong their lives.

Even a good filter however, may let through 3 out of 100 000 organisms, so when hazardous organisms are being worked with and the exhaust air is vented into areas frequented by people two filters, in series, are used.

5.10 Treatment of waste material

It is the responsibility of all laboratory workers to ensure that no infected material should ever leave the laboratory and become a risk to other workers or to the general public. Such material should be treated to render it safe. The practical methods of treating contaminated laboratory waste are:

- chemical disinfection;
- sterilisation by autoclaving;
- sterilisation by incineration.

The choice of the method must be determined by the nature of the material to be treated. Figure 5.6 shows possible routes for the treatment of infected material.

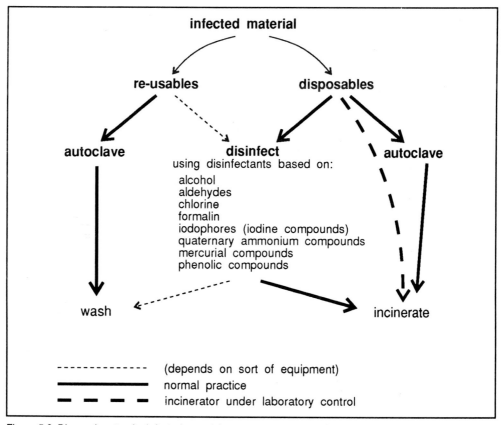

Figure 5.6 Disposal routes for infected material.

5.10.1 Disinfection

For general and routine use, the preferred disinfectant is a solution of sodium hypochlorite. This should be used in dilutions made from stock solutions using the manufacturers' recommendations.

a) Spill control

A concentration of 10,000 ppm available chlorine (for example 1/10 dilution of "neat" Chloros) is recommended for applying to small spillages on the bench prior to mopping-up. "Squeezy" plastic bottles containing this concentration of hypochlorite should be readily available and the hypochlorite should be replaced every two days.

b) Disinfection of re-usable pipettes

Pipettes discarded into disinfectant solution for decontamination must be completely immersed. The recommended concentration of hypochlorite solution for this is 2500 ppm. Pipettes must be immersed overnight and they should be removed from the disinfectant first thing in the morning, before the start of laboratory work. The hypochlorite soultion in the pipette discard jars must be replaced at least twice weekly.

c) Routine bench swabbing

hypochlorite

When work is finished at the bench, it is a good precaution to swab the bench with a weak solution of sodium hypochlorite. Usually, a concentration of 1000 ppm is adequate for this purpose.

phenolic
disinfectants

All containers of hypochlorite solution should be clearly marked with the concentration and the use for which that particular dilution has been designed. Sodium hypochlorite is a strong oxidising agent and thus a powerful disinfectant. It is ideal for glassware which is going to be washed and re-used. However, its efficacy is much affected with prolonged exposure due to evaporation of chlorine and by reduction of the hypochlorite. For these reasons, phenolic disinfectants are prefered for the disposal of non re-usable items such as microscope slides and Pasteur pipettes. Proprietary disinfectants such as Lysol or Clearsol are used at a concentration of 1% and should be renewed at regular intervals, usually twice weekly. However, phenolic disinfectants are not recommended for re-usable glassware as it is difficult, if not impossible, for all traces of the disinfectant to be removed from the glass surface. In laboratories growing animal cells and virology laboratories, hypochlorite rather than phenolic disinfectants should be used for all chemical disinfections whether or not the items to be disinfected are to be re-used.

5.10.2 Autoclave

Material for autoclaving should be transported to the autoclave in robust, leakproof containers of a design approved by the Biological Safety Officer for the laboratory. Such containers may be constructed of either metal or rigid plastic which will withstand autoclave temperatures. Ideally, these containers would be fitted with a removable lid. The autoclave should be tested regularly with thermocouple probes to ensure that the correct temperature is being achieved throughout the chamber. If portable bench autoclaves are used, they should be fitted with timer devices as an additional safety precaution. At the end of each working day, all cultures should be removed from open benches. Cultures to be kept at room temperature should be transported to a secure place.

5.11 Routine laboratory procedures associated with hazard

5.11.1 Ultrasonic disruption of fluid material

The use of an ultrasonic probe may be particularly hazardous. Here, the probe is inserted into the material to be disrupted, usually through the cap of an universal container. Inevitably, this will generate an aerosol. The small droplets of an aerosol dry rapidly and may become droplet nuclei which can remain in the atmosphere for several hours. An additional problem with this technique is the localised heat generated at the tip of the probe. Disrupting fluids in a sealed bottle immersed in the water of an ultrasonic cleaning bath is far preferable to using a probe, for most purposes. This technique has the additional advantage that the material being disrupted suffers only a marginal rise in temperature. Care has to be taken to ensure that the bottle used for holding the specimen in the ultrasonic bath is strong enough for the task - i.e. that the bottom of the bottle is not likely to fall off during the procedure, thus creating something of a problem!

5.11.2 The use of a wire loop

The withdrawal of a loop from a liquid culture will create droplets as will a film of culture on a loop when it bursts. When a loop, charged with culture, is placed in a flame it will sputter, liberating droplets which have not been heat-sterilized. Many laboratories attach sputter guards to Bunsen burners to overcome this problem.

5.11.3 Laboratory note-books

Many people absent-mindedly suck the end of pencils and pens. If such implements are placed on the laboratory bench, then contamination may be introduced into the mouth. It is essential that pens and pencils are placed on the note-book rather than on the bench and that the note book is kept well away from experimental work.

5.11.4 Centrifugation

If sensible precautions are followed, centifugation should not be a particularly hazardous operation. In bench centrifuges, screw-topped containers should be used wherever possible. The containers should be balanced and the bottom of the centrifuge bucket should have a rubber cushion to prevent the problem of glass tubes cracking with centrifugal force. The centrifuge lid must never be lifted until the rotor has stopped spinning. Ultracentifuges should have a disinfection trap on the exhaust ventilation in case of contamination within the bowl of the centrifuge.

5.11.5 Incubation and storage of cultures

All culture bottles and Petri dishes should be stored in secure racks in both incubators and refrigerators. Many laboratory accidents are caused by insecure cultures being knocked over. Also, it is imperative that all cultures are clearly labelled (including the name of the experimentor).

Always remember that there is no such thing as a completely safe micro-organism. Organisms non-pathogenic for animals may be severly pathogenic for plants for example. Also, organisms which inhabit, for example the skin, may not cause disease in the concentration in which they occur naturally, but in the large numbers seen in laboratory cultures they could pose a real threat to health.

5.12 The special issues of animal cell cultures

So far we have predominantly discussed the safety issues concerned with the culture of micro-organisms. It is, however, important to consider animal cell cultures since they may present problems of safety. Currently animal cell lines are used to produce vaccines and many 'immortalised' (hybridoma) cell lines are used to produce monoclonal antibodies. Increasingly cell lines such as Chinese Hamster Ovary (CHO) cells are used to produce a variety of human and other mammalian proteins. These cell lines particularly lend themselves to genetic manipulation.

We will not review the origins of these cell lines in detail here (this aspect is covered in the BIOTOL text, 'In vitro Cultivation of Animal Cells'). We will, however, give a brief review of the major stages in producing cell lines as these have some bearing on the safety issues associated with such cell lines.

5.12.1 Methods used in the culture of animal cells

Initially animal cell cultures were of much importance to the study of how viruses attack human and animal cells. The technique of growing cells in vitro involves the use of animal cells from different organs. Generally, pieces of tissue are treated with trypsin solution to obtain single cells. Then a drop of this suspension is placed onto a flat surface (for example, in a Petri dish or on the inside of a flattened bottle of flask). The cells adhere to the surface and will grow if supplied with a nutrient solution containing about 10 amino acids and 10 growth factors or vitamins as well as salts, glucose and bicarbonate solution in equilibrium with 5% CO_2 in the gas phase. Often 5% calf or foetal calf serum is added.

primary culture   At this stage, the cells of this **primary culture** may still be contaminated with bacteria and viruses. Viruses can be detected by abnormalities in the growth of the cells and antibiotics such as penicillin and streptomycin are added to stop bacterial growth.

The cells will multiply until they occupy the whole surface (confluent cells). A few cells are then transferred to a sterile bottle with fresh sterile medium. Note that the cells only grow as a monolayer on the surface of the culture vessel.

The overall properties of the cells generally remain the same over many generations. Mutations may, however, occur and a particular mutant may become dominant.

established cell lines   These cells derived from normal cells cannot be subcultured indefinitely. After some 40-50 transfers the growth rate drops and the cells change their appearance: from the characteristic, normal euploid (diploid) pattern to an aneuploid stage, whereby numerous chromosomes and defective chromosome fragments appear. Then the cells die. Malignant tissue cells, however, give rise directly to cell lines that have an indefinite life span. If they have been grown for over 40 subcultures they may be considered as **established cell lines**. Malignant cells are often able to grow in suspension and much higher cell densities can be achieved than can be produced with normal cells.

vaccine production   Human viral vaccines may be produced by growing viruses on normal cells such as human embryo fibroblasts with finite life and non-tumorigenic characteristics. However, only limited quantities of vaccine can be produced (there are exceptions, for example, the large-scale production of Foot-and-Mouth Disease vaccine with the aid of baby hamster kidney cells).

Since malignant cells are able to grow in suspension at quite high densities, and are 'immortalised' (able to be sub-cultured indefinitely). Increasing use is made of these cell lines for vaccine production. In this way unlimited quantities of antigens may be produced.

Antigen-specific hybridomas and monoclonal antibodies are obtained by fusing myeloma (tumour) cells with lymphocytes of for instance, mice, which have been repeatedly immunised with an antigen of interest. Hybridomas may be grown in suspension for the production of monoclonals antibodies for commercial diagnostic uses.

In summary, animal cell cultures play an important role in the production of:

- human vaccines: (Polio, Measles, Mumps, Rubella, etc);

- veterinary vaccines: (Foot-and-Mouth Disease, Rabies);

- monoclonal antibodies: (for immune diagnosis, pregnancy tests, diagnosis of tumours, HIV tests, etc);

- human proteins: (for example erythropoietin).

5.12.2 Safety considerations in using animal cells

oncogenes

Health complications have not been reported in laboratory work during 30 years of experience with cell lines. Nevertheless, animal cells contain proto-oncogene sequences in the cellular DNA. These can, in principle, be activated spontaneously under certain conditions and become carcinogenic or tumorigenic (eg by tumorigenic substances or retrovirus infections). Transformed cells may cause transplantable tumours in sensitive test animals but very seldom in humans in the case of accidental injury. Also remember that animal cells may harbour latent viruses.

latent viruses

The recommendations, therefore, concerning the safe handling of cell cultures (tissue explants, primary cell cultures, continuous cell lines) of human and animal origin as well as cell products derived from them, are also concerned with activated oncogenes and the possibility that such cells/cell products may harbour latent viruses.

contamination from other sourses

In addition explants and primary cell cultures may be contaminated with micro-organisms originating from human or animal donors, eg viruses and mollicutes (formerly mycoplasma). Furthermore, cell cultures may be contaminated accidentally during manipulations or via contaminated media components such as sera. It is, therefore, useful to list recommendations for the safe handling of animal cell culture.

5.12.3 Risks and safety precautions inherent with work with animal cells and cell lines

Primary cell cultures

Since the origin of microbiological contaminations in primary cell cultures are not known, all manipulations involving primary cell cultures should take place under Containment Category 1 (UK Hazard Group 2, Netherlands PG2) conditions, **unless** there are indications that medium-risk (EFB3) or high-risk organisms (EFB4) are present as contaminants. In these cases, the work should be performed under Containment Category 2 (CC2) or Containment Category 3 (CC3) conditions respectively. This may be the case with cells derived from primates with possible retrovirus infections; CC3 is then indicated. Retroviruses may be amphotrophic, that is, they may grow in related species.

The presence of pathogenic contaminants can be excluded if the cells were derived from healthy donors of non-suspect origin, eg germ-free (gnotobiotic) laboratory animals. Handling does not require physical containment in such cases.

Cells of specific pathogen-free (SPF) laboratory animals may be used. Physical containment is then only dependent on the type of micro-organisms which may be still be present.

Established cell lines

All established cell lines should be considered as having potential low-risk (EFB2) and should consequently be handled under CC1 conditions, unless the cell lines are suspected of being contaminated with medium- or high-risk organisms. In these cases, the work should be carried out under CC2 or CC3 conditions. However, if the presence of pathogens in established tumour cell lines can be excluded, the cell line can be handled under GILSP conditions.

5.12.4 Risks involved with cell products

The risks connected with work with cell products from animal cell cultures (eg supernatants of cells, ascites and amnion fluids) may be classified as described below.

Cell products containing cells.

They should be handled according to the physical containment prescribed for the cells or cell lines from which they are derived.

Cell-free products

When cells are absent and, therefore, only microbiological contaminations may play a role, the following physical containment levels should be considered:

- cell-free products which do not carry any microbiological contaminants;
- no physical containment is required, eg for mouse monoclonal antibodies in the absence of human pathogens;
- cell-free products which may be contaminated with a known pathogenic micro-organism;
- these cell-free cellular products should be handled under the same physical containment as the contaminating micro-organism.

When the pathogen has been eliminated from the products or inactivated, no physical containment for further handling is required. It is necessary to have the elimination or inactivation validated by standardised methods for every batch, for instance:

chemical inactivation:

standard determination of the residual concentration of the inactivating agent employed;

physical inactivation of retroviridae:

determination of the absence of reverse transcriptase activity.

We have summarised the risk class and safety precautions to be taken with animal cells, established cell lines and cell products in Table 5.8.

Animal cells/cell products	Risk class and safety precautions	
	Laboratory	Pilot plant
primary cell cultures		
unknown contaminants	low-risk (EFB2)	Containment Category 1
contains known pathogens	according to classification of pathogen	
pathogens are absent	harmless (EFB1)	GILSP
established cell lines/tumour cell lines		
unknown contaminants	low-risk (EFB2)	Containment Category 1
contains known pathogens	according to classification of pathogen	
pathogens are absent	harmless (EFB1)	GILSP
cell products		
contains viable cells	according to risk class of cell culture	
cell-free:		
original culture free from pathogens	harmless (EFB1)	GILSP
original culture with known pathogens	according to classification of pathogen	
validated and controlled inactivation of known pathogen	harmless (EFB1)	GILSP

Table 5.8 Summary of the risk classes of primary and established animal cell lines. EFB1 = European Federation of Biotechnology, Class 1 = harmless, GILSP = good industrial large-scale practice.

Appendix 5.1 - Categorisation of micro-organisms according to hazard

Within the main body of the text, we have referred to the fact that micro-organisms may be categorised according to the extent that they represent a hazard to health. Four groups are identified - hazard group 1 being of lowest risk, hazard group 4 presenting the greatest hazard.

The extent and nature of the containment that needs to be implemented when using micro-organisms is governed to a large extent by the hazard group to which the micro-organism belongs.

Here we provide a list of organisms according to their hazard group. The list includes only organisms in hazard groups 2, 3 and 4. The names that have been used are those in common (but not universal) usage. Many have one or more synonyms.

The list provided here is based on the categorisation of pathogens by the UK, Advisory Committee on Pathogens ('Categorisation of Pathogens according to hazard and categories of containment' HMSO, 1990).

In countries other than the UK, a slightly different nomenclature may be used although the same (or very similar) criteria for assigning micro-organisms to 1 of 4 categories is often employed.

The list provided here should not be regarded as universally applicable for all time and for all places. As new knowledge emerges or as new control measures are developed, it becomes appropriate to re-categorise some organisms. Furthermore, it is possible to develop attenuated or genetically modify strains of pathogenic organisms such that they may be re-categorised.

The reader should also be aware that specific conditions may apply to particular pathogens. For example, deliberate cultivation of the Variola (smallpox) virus is banned totally in some countries. A key aspect of working with micro-organisms is therefore, not only to work safely, but also to be aware of the regulatory obligations that must be fulfilled.

BACTERIA, CHLAMYDIAS, RICKETTSIAS AND MYCOPLASMAS

Hazard Group 3

Bacillus anthracis

Brucella spp.

Chlamydia psittaci (avian strains only)

Coxiella burnetti

Francisella tularensis (Type A)

Mycobacterium africanum

Mycobacterium avium

Mycobacterium bovis (excl BCG strain)

Mycobacterium intracellulare

Mycobacterium kansasii

Mycobacterium leprae

Mycobacterium malmoense

Mycobacterium paratuberculosis

Mycobacterium scrofulaceum

Mycobacterium simiae

Mycobacterium szulgai

Mycobacterium tuberculosis

Mycobacterium xenopi

Pseudomonas mallei

Pseudomonas pseudomallei

Rickettsia-like organisms

Rickettsia spp.

Salmonella paratyphi A, B, C

Salmonella typhi

Shigella dysenteriae (Type 1)

Yersinia pseudotuberculosis subsp pestis (Y pestis)

Hazard Group 2

Acinetobacter calcoaceticus

Acintoebacter lwoffi

Actinobacillus spp.

Actinomadura spp.

Actinomyces bovis

Actinomyces israelii

Aeromonas hydrophila

Alcaligenes spp.

Arizona spp.

Bacillus cereus

Bacteroides spp.

Bacterionemia matruchottii

Bartonella bacilliformis

Bordetella parapertussis

Bordetella pertusis

Borrelia spp.

Campylobacter spp.

Cardiobacterium hominis

Chlamydia spp. (other than aivan strains)

Clostridium botulinum

Legionella spp.

Leptospira spp.

Listeria monocytogenes

Moraxella spp.

Morganella morganii

Mycobacterium bovis (BCG strain)

Mycobacterium chelonei

Mycobacterium fortuitum

Mycobacterium marinum

Mycobacterium microti

Mycobacterium ulcerans

Mycoplasma pneumoniae

Neisseria spp. (spp .known to be pathogenic for man)

Nocardia asteroides

Nocardia brasiliensis

Pasteurella spp.

Peptostreptococcus spp.

Plesiomonas shigelloides

Proteus spp.

Providencia spp.

Hazard Group 2 (Cont)

Clostridium tetani

Clostridium spp. (other spp. known to be pathogenic for man)

Corynebacterium diphtheriae

Corynebacterium spp. (other spp. known to be pathogenic for man)

Edwardsiella tarda

Eikenella corrodens

Enterobacter spp.

Erysipelothrix rhusiopathiae

Escherichia coli (except those known to be non-pathogenic)

Flavobacterium meningosepticum

Francisella tularensis (Type B)

Fusobacterium spp.

Gardnerella vaginalis

Haemophilus spp.

Hafnia alvei

Kingella kingae

Klebsiella spp.

Pseudomonas spp. (other spp. known to be pathogenic for man)

Salmonella spp. (other than those in Hazard Group 3)

Serratia liquefaciens

Serratia marcescens

Shigella spp. (other than that in Hazard Group 3)

Staphylococcus aureus

Streptobacillus moniliformis

Streptococcus spp. (except those known to be non-pathogenic for man)

Treponema pallidum

Treponema pertenue

Veillonella spp.

Vibrio cholerae (incl El Tor)

Vibrio parahaemolyticus

Vibrio spp. (other species known to be pathogenic for man)

Yersinia enterocolitica

Yersinia pseudotuberculosis subsp pseudotuberculosis

FUNGI

Hazard Group 3

Blastomyces dermatitidis

(Ajellomyces dermatitidis)

Coccidioides immitis

Histoplasma capsulatum var. capsulatum

(Ajellomyces capsulata)

Histoplasma capsulatum var duboisii

Histoplasma capsulatum var farciminosum

Paracoccidioides brasiliensis

Penicillium marneffei

Hazard Group 2

Absidia corymbifera

Acremonium falciforme

Acremonium kiliense

Acremonium recifei

Aspergillus flavus

Aspergillus fumigatus

Aspergillus nidulans

Exophialia jeanselmei

Exophialia spinifera

Exophialia richardsiae

Fonsecaea compacta

Fonsecaea pedrosoi

Fusarium solani

Fusarium oxysporum

Hazard Group 2 (Cont)

Aspergillus niger	*Geotrichum candidum*
Aspergillus terreus	*Hendersonula toruloidea*
Basidiobolus haptosporus	*Leptosphaeria senegalensis*
Candida albicans	*Madurella mycetomatis*
Candida glabrata	*Madurella grisea*
Candida guilliermondii	*Malassezia furfur*
Candida krusei	*Microsporum spp.*
Candida parapsilosis	*Neotestudina rosatii*
Candida kefyr	*Phialophora verrucosa*
Candida tropicalis	*Piedraia hortae*
Cladosporium carrionii	*Pneumocytis carinii*
Conidiobolus coronatus	*Pseudallescheria boydii*
Cryptococcus neoformans	*Pyrenochaeta romeroi*
(Filobasidiella neoformans)	*Rhizomucor pusillus*
Cunninghamella elegans	*Rhizopus microsporus*
Curvularia lunata	*Rhizopus oryzae*
Emmonsia parva	*Sporothrix schenckii*
Emmonsia parva var. crescens	*Trichophyton spp.*
Epidermophyton floccosum	*Trichosporon beigelii*
Exophialia dermitidis	*Xylohypha bantiana*
Exophialia werneckii	

PARASITES

Hazard Group 3

Echinococcus spp.	*Toxoplasma gondii*
Leishmania spp. (mammalian)	*Trypanosoma cruzi*
Naegleria spp.	

Hazard Group 2

Acanthamoeba spp.	*Loa loa*
Ancylostoma duodenale	*Mansonella ozzardi*
Angiostrongylus spp.	*Necator americanus*
Ascaris lumbricoides	*Onchocerca volvulus*
Babesia microti	*Opisthorchis spp.*
Babesia divergens	*Paragonimus westermanni*
Balantidium coli	*Plasmodium spp.* (human & simian)
Brugia spp.	*Pneumocystis carinii*
Capillaria spp.	*Schistosoma haematobium*

Hazard Group 2 (Cont)

Clonorchis sinensis

Cryptosporidium spp.

Dipetalonema streptocerca

Dipetalonema perstans

Diphyllobothrium latum

Drancunculus medinensis

Entamoeba histolytica

Fasciola hepatica

Fasciola gigantea

Fasciolopsis buski

Giardia lamblia

Hymenolepis nana (human origin)

Hymenolepis diminuta

Schistosoma intercalatum

Schistosoma japonicum

Schistosoma mansoni

Stronglyloides spp.

Taenia saginata

Taenia solium

Toxocara canis

Trichinella spp.

Trichomonas vaginalis

Trichostrongylus spp.

Trichuris trichiura

Trypanosoma brucei subsp

Wuchereria bancrofttii

VIRUSES

Hazard Group 4

Arenaviridae

Junin virus

Lassa fever virus

Machupo virus

Mopeia virus

Bunyaviridae

Nairoviruses

Congo/Crimean haemorrhagic fever

Togaviridae

Flaviviruses

Tick-borne viruses

Filoviridae

Ebola virus

Marburg virus

Absettarov

Hanzalova

Hypr

Poxviridae

Variola (major & minor) virus

('whitepox' virus)

Kyasanur Forest

Omsk

Russian spring-summer encephalitis

Hazard Group 3

Arenaviridae

Lymphocytic choriomeningitis virus (LCM)

Rhabdoviridae

Rabies virus

Hazard Group 3 (Cont)

Bunyaviridae

Bunyamwera supergroup

Oropouche virus

Phleboviruses

Rift Valley fever

Hantaviruses

Hantaan (Korean haemorrhagic fever)

Other hantaviruses

Hepadnaviridae

Hepatitis B virus

Hepatitis B virus + Delta

Herpesviridae

Herpesvirus simiae (B virus)

Poxviridae

Monkeypox virus

Retroviridae

Human immunodeficiency viruses (HIV)

Human T-cell lymphotropic viruses (HTLV)
types 1 and 2

Hazard Group 2

Adenoviridae

Arenaviridae

other arenaviruses

Astroviridae

Bunyaviridae

Hazara virus

other bunyaviruses

Caliciviridae

Coronaviridae

Herpesviridae

Cytomegalovirus

Epstein-Barr virus

Herpes simplex viruses types 1 and 2

Herpesvirus varicella-zoster

Human B-lymphotropic virus (HBLV - human
herpesvirus type 6)

Togaviridae

Alphaviruses

Eastern equine
encephalomyelitis

Venezuelan equine
encephalomyelitis

Western equine
encephalomyelitis

Flaviviruses

Japanese B encephalitis

Kumlinge

Louping ill

Murray Valley encephalitis (Australia
encephalitis)

Powassan

Rocio

St Louis encephalitis

Tick-borne encephalitis

Yellow fever

Picornaviridae

Acute haemorrhagic conjunctivitis virus (AHC)

Coxsackieviruses

Echoviruses

Hepatitis A virus (human enterovirus type 72)

Polioviruses

Rhinoviruses

Poxviridae

Cowpox virus

Molluscum contagiosum virus

Orf virus

Vaccinia virus

Reoviridae

Human rotaviruses

Orbiviruses

Reoviruses

Orthomyxoviridae

Influenza viruses types A, B & C

Influenza virus type A-recent isolates

Paramyxoviridae

Measles virus

Mumps virus

Newcastle disease virus

Parainfluenza viruses types 1 to 4

Respiratory syncytial virus

Papovaviridae

BK and JC viruses

Human papillomaviruses

Parvoviridae

Human parvovirus (B19)

Rhabdoviridae

Vesicular stomatitis virus

Togaviridae

Other alphaviruses

Other flaviviruses

Rubivirus (rubella)

Unclassified viruses

Hepatitis non-A non-B viruses

Norwalk-like group of small round structured viruses

Small round viruses (SRV - associated with gastroenteritis)

Unconventional agents associated with:

Creutzfeldt-Jakob disease

Gertsmann-Sträussler-Schienker syndrome

Kuru

Safety and the genetic manipulation of organisms

Safety and the genetic manipulation of organisms

6.1 Introduction

The undoubted advantages that may be accrued from genetically manipulating organisms are, to some extent, counteracted by the potential hazards which may arise from carrying out genetic manipulations. We may divide these hazards into two groups:

- those which threaten those who handle genetically modified organisms or carry out genetic manipulating procedures;

- those which offer a threat to the public at large.

The former of these two groups can be considered as the hazards associated with the contained uses of genetically manipulated organisms. The latter group are mainly those hazards associated with the release of manipulated organisms.

This text is primarily concerned with the laboratory procedures for genetically manipulating organisms, especially micro-organisms.

From the onset you must realise that the contained use of genetically manipulated systems is the subject of both international and national regulations. Currently there are no universally accepted rules governing the genetic manipulation and use of micro-organisms although there are many similarities between the various regulations.
EC directives For example the EC has produced two directives (OJ, L117, Vol 33, 8th May, 1990). One of these is concerned with the contained (laboratory) use of genetically manipulated organisms, the other specifies procedures for the deliberate release of genetically modified micro-organisms.

We will begin by providing a summary of the EC-Directive on the contained use of genetically modified micro-organisms (EC-Directive 90/219/EEC). We will then consider the regulations and practices in Member States. For this, we will particularly draw upon the procedures operating within the UK and the Netherlands. We will then provide a summary of the EC-Directive on the procedures used for the deliberate release of genetically modified organisms. We will complete the chapter by considering the issues that need to be included in determining the risks in releasing genetically modified organisms.

licensing In The Netherlands all recombinant DNA projects are regulated by the Nuisance Act.
arrangements Facilities in which such work is carried out require a licence. The licence is given by the
Commission Community Council and is based on the advice of the Commission Genetic Modication
Genetic (CGM). The Commission classifies each project into a risk group and advises on the
Modification safety measures to be implemented. A parallel arrangement operates in the UK. In the
(CGM) and latter case, the Advisory Committee on Genetic Manipulation (ACGM) fulfils a similar
ACGM function to the Commission Genetic Modification in The Netherlands.

The safety conditions of the workers in laboratories and industries involved in biotechnology are regulated by the ARBO law (Arbeidsomstandighederwet) in The Netherlands whilst in the UK this falls within the Health and Safety at Work Act. In both cases, the responsibility of the director of the institute or industry regarding the protection of workers is specified. In both cases, a Safety Officer with special training in recombinant DNA safety must be appointed. In The Netherlands the position carries the title Biosafety Officer (BSO) whilst in the UK the usual title is Biological Safety Officer.

In both countries, each project involving recombinant DNA has to be risk assessed and placed into a risk (hazard) category. This assessment is usually first carried out in-house by those wishing to carry out the project and the outcome of this assessment is reported to the appropriate body (eg CGM in The Netherlands, ACGM in the UK). These bodies judge each proposal on merit and may advise granting approval of the project. Other countries, especially EC states and the USA, have analogous arrangements.

6.2 The EC directive on the contained use of genetically modified micro-organisms (90/219/EEC)

The EC Council Directive of 23rd April, 1990 on the contained use of genetically modified organisms is published in the Official Journal of the European Communities, L117, Volume 33, 8th May, 1990. The purpose of the directive is to 'lay down common measures for the contained use of genetically modified micro-organisms with a view to protecting human health and the environment'. Progressively, national regulations are being adjusted to fit with these supra-national directives. In common with many such documents the legal wording of the EC Directives can be somewhat off-putting! The following sections may therefore be a little difficult but are worth persevering with. In this and other directives, the definitions which are used and the exemptions to the directive are of great importance to its scope and meaning.

Our first task is therefore to consider these.

6.2.1 Definitions

A **micro-organism** is defined as 'any microbial entity, cellular or non-cellular, capable of replication or transferring genetic material'.

A **genetically modified micro-organism**, which we shall abbreviate to GMMO, is defined as 'a micro-organism in which the genetic material has been altered in a way that does not occur naturally by mating and/or natural recombination'.

In order to indicate techniques of genetic modification the following (non-exhaustive) list is given:

- recombinant DNA techniques using vector systems;

- techniques involving the direct introduction into a micro-organism of heritable material prepared outside the micro-organism including micro-injection, macro-injection and micro-encapsulation;

- cell fusion or hybridisation techniques to form new combinations of heritable genetic material which do not occur naturally.

Contained use means 'any operation in which micro-organisms are genetically modified or in which such organisms are cultured, stored, used, transported, destroyed or disposed of and for which physical barriers together with chemical and/or biological barriers, are used to limit their contact with the general population and the environment.

As we have seen, definitions are of great importance. Exemptions are also of great importance in regulatory affairs. They can on occasion make the difference between an experiment or procedure being subject to a regulation or not.

6.2.2 Exemptions

The following techniques are said not to result in genetic modification on condition that they do not involve the use of recombinant DNA or genetically modified organisms:

- *in vitro* fertilisation;
- conjugation, transduction, transformation or any other natural process;
- polyploidy induction.

The techniques of genetic manipulation which are excluded from the Directive so long as they do not involve the use of genetically modified micro-organisms as recipient or parental organisms are the following:

- mutagenesis;
- construction and use of somatic animal hybridoma cells;
- cell fusion of cells from plants which can be produced by traditional breeding methods;
- self-cloning of non-pathogenic naturally occurring micro-organisms which fulfil the criteria of Group 1 (see below) for recipient organisms.

Furthermore this directive does not apply to the transport of GMMOs or the storage, transport, destruction or disposal of GMMOs which have been placed on the market under community legislation, which includes an appropriate specific risk assessment.

At the time of writing, the applicability of some of these issues to the UK and The Netherlands is still under discussion.

6.2.3 Classification of GMMOs within the directive

For the purposes of the Directive, GMMOs must be classified into two groups, Group I and Group II.

Group I

Group I organisms have a long history of safe use and are considered to be safe when used under specific conditions. These criteria are similar in many ways to those we discussed in Chapter 5 (see Table 6.1).

Criteria for classifying genetically modified micro-organisms in Group I

A   **Recipient or parental organism**

—   non-pathogenic;
—   no adventitious agents;
—   proven and extended history of safe use or built-in biological barriers, which, without interfering with optimal growth in the reactor or fermenter, confer limited survivability and replicability, without adverse consequences in the environment.

B   **Vector/insert**

—   well characterised and free from known harmful sequences;
—   limited in size as much as possible to the genetic sequences required to perform the intended function;
—   should not increase the stability of the construct in the environment (unless that is a requirement of intended function);
—   should be poorly mobilisable;
—   should not transfer any resistance markers to micro-organisms not known to acquire them naturally (if such acquisition could compromise use of drug to control disease agents).

C   **Genetically modified micro-organisms**

—   non-pathogenic;
—   as safe in the reactor or fermenter as a recipient or parental organism, but with limited survivability and/or replicability without adverse consequences in the environment.

D   **Other genetically modified micro-organisms that could be included in Group I if they meet the conditions in C above.**

—   those constructed entirely from a single prokaryotic recipient (including its indigenous plasmids and viruses) or from a single eukaryotic recipient (including its chloroplasts, mitochondria, plasmids, but excluding viruses);
—   those that consist entirely of genetic sequences from different species that exchange these sequences by known physiological processes.

Table 6.1 Annex II from EC Council Directive of 23rd April, 1990 on the use of genetically modified micro-organisms.

Group II

Those GMMOs which are not in Group I, are said to fall into Group II.

Broadly speaking EFB Class 1, the Dutch PG Group 1 and the UK Hazard Group 1, which we discussed in Chapter 5 can be compared with Group I and the other EFB, PG and Hazard Groups as Group II.

Having classified a GMMO into the category Group I or II, the next step is to consider the type of operation or manipulation which is to be applied to the GMMO.

Types of operation

Two types of operation are considered in the 1990 EC Directive on the use of genetically manipulated organisms known as Type A and Type B operations.

Type A operations are 'any operation used for teaching, research, development, or non-industrial or non-commercial purposes and which is of a small scale (eg 10 litres or less)'.

Type B operations are any operations other than Type A operations.

It should be noted that for Type A operations some of the criteria shown in Annex II may not be applicable in determining the classification of all GMMOs. In which instance a competent authority should ensure that the relevant criteria are met.

6.2.4 The system of the Directive

The Directive states that all appropriate measures should be taken to avoid all adverse effects on human health and the environment through the contained use of GMMOs. To this end, the user is obliged to carry out a prior assessment of the risks that may occur taking into account the parameters which are relevant from Annex III of the Directive. Annex III is shown in Table 6.2.

A record of this assessment should be kept by the user and made available to the competent authority if appropriate.

Safety assessment parameters to be taken into account, as far as they are relevant, in accordance with Article 6 (3)

A Characteristics of the donor, recipient or (where appropriate) parental organism(s).

B Characteristics of the modified micro-organism

C Health considerations

D Environmental considerations

A **Characteristics of the donor, recipient or (where appropriate) parental organism(s)**

— names and designation;
— degree of relatedness;
— sources of the organism(s);
— information on reproductive cycles (sexual/asexual) of the parental organism(s) or, where applicable, of the recipient micro-organism;
— history of prior genetic manipulations;
— stability of parental or of recipient organism in terms of relevant genetic traits;
— nature of pathogenicity and virulence, infectivity, toxicity and vectors of disease transmission;
— nature of indigenous vectors:
  sequence,
  frequency of mobilisation,
  specificity,
  presence of genes which confer resistance;
— host range;
— other potentially significant physiological traits;
— stability of these traits;
— natural habitat and geographical distribution. Climatic characteristics of original habitats;
— significant involvement in environmental processes (such as nitrogen fixation or pH regulation);
— interaction with, and effects on, other organisms in the environment (including likely competitive or symbiotic properties);
  ability to form survival structures (such as spores or sclerotia).

Table 6.2 Annex III from EC Council Directive of 23rd April, 1990 on the use of genetically modified micro-organisms.   (Continued).

B   **Characteristics of the modified micro-organism**

— the description of the modification including the method for introducing the vector-insert into the recipient organism or the method used for achieving the genetic modification involved;
— the function of the genetic manipulation and/or of the new nucleic acid;
— nature and source of the vector;
— structure and amount of any vector and/or donor nucleic acid remaining in the final construction of the modified micro-organism;
— stability of the micro-organism in terms of genetic traits;
— frequency of mobilisation of inserted vector and/or genetic transfer capability;
— rate and level of expression of the new genetic material. Method and sensitivity of measurement;
— activity of the expressed protein.

C   **Health considerations**

— toxic or allergenic effects of non-viable organisms and/or their metabolic
— products;
— product hazards;
     comparison of the modified micro-organism to the donor, recipient or (where
— appropriate) parental organism regarding pathogenicity;
— capacity for colonisation;
     if the micro-organism is pathogenic to humans who are immunocompetent:
     a) diseases caused and mechanisms of pathogenicity including invasiveness and   virulence;
     b) communicability;
     c) infective dose;
     d) host range, possibility of alteration;
     e) possibility of survival outside of human host;
     f) presence of vectors or means of dissemination;
     g) biological stability;
     h) antibiotic-resistance patterns;
     i) allergenicity;
     j) availability of appropriate therapies.

D   **Environmental considerations**

— factors affecting survival, multiplication and disseminations of the modified micro-organism in the environment;
— available techniques for detection, identification and monitoring of the modified micro-organism;
— available techniques for detecting transfer of the new genetic material to other organisms;
— known and predicted habitats of the modified micro-organism;
— description of ecosystems to which the micro-organism could be accidentally disseminated;
— anticipated mechanism and result of interaction between the modified micro-organism and the organisms or micro-organisms which might be exposed in case of release into the the environment;
— known or predicted effects on plants and animals such as pathogenicity, infectivity, toxicity, virulence, vector of pathogen, allergenicity, colonisation;
— known or predicted involvement in biogeochemical processes;
— availability of methods for decontamination of the area in case of release to the environment.

Table 6.2 Annex III from EC Council Directive of 23rd April, 1990 on the use of genetically modified micro-organisms.

6.2.5 Containment requirements for Group I and II GMMOs
Containment requirements for Group I GMMOs

The containment requirements for Group I organisms are set out in Article 7 of the Directive which is shown in Table 6.3.

For genetically modified micro-organisms in Group I, principles of good microbiological practice, and the following principles of good occupational safety and hygiene, shall apply:

i)	to keep workplace and environmental exposure to any physical, chemical or biological agent to the lowest practicable level;
ii)	to exercise engineering control measures at source and to supplement these with appropriate personal protective clothing and equipment where necessary;
iii)	to test adequately and maintain control measures and equipment;
iv)	to test, when necessary, for the presence of viable process organisms outside the primary physical containment;
v)	to provide training of personnel;
vi)	to establish biological safety committees or subcommittees as required;
vii)	to formulate and implement local codes of practice for the safety of personnel.

Table 6.3   Article 7 of the EC Council Directive of 23rd April, 1990 on the use of genetically modified micro-organisms.

Containment requirements for Group II GMMOs

In addition to Article 7 other containment measures should be applied to Group II GMMOs; they are set out in Annex IV of the Directive, which is shown in Table 6.4.

6.2.6 Activities which require authorisation or notification

For the purposes of the directive, the two Groups of GMMOs and the two types of operations can be split into four categories.

- Type A operations with Group I organisms - IA operations;

- Types B operations with Group I organisms - IB operations;

- Type A operations with Group II organisms - IIA operations;

- Type B operations with Group II organisms - IIB operations.

first use requires notification

The first use of an installation for an operation involving GMMOs is considered to be an activity for which notification of the authorities is required prior to commencement of the work.

conditions apply for subsequent use

Thereafter, Articles 9, 10 and 11 and Annex V of the Directive specify the type of notification which is required for each of the four categories of operations. They also describe the time lengths which in the case of Group IB and IIA operations must be waited before, in the absence of any indication to the contrary from the authorities, work can commence, or, in the case of Group IIB operations, the maximum length of time which the authorities will take to communicate their decision on whether or not the work may proceed.

Containment measures for micro-organisms in Group II

The containment measures for micro-organisms from Group II shall be chosen by the user from the categories below as appropriate to the micro-organisms and the operation in question in order to ensure the protection of the public health of the general population and the environment.

Type B operations shall be considered in terms of their unit operations. The characteristics of each operation will dictate the physical containment to be used at that stage. This will allow selection and design of process, plant and operating procedures best fitted to assure adequate and safe containment. Two important factors to be considered when selecting the equipment needed to implement the containment are the risk of, and the effects consequent on, equipment failure. Engineering practice may require increasingly stringent standards to reduce the risk of failure as the consequence of that failure becomes less tolerable.

Specific containment measures for Type A operations shall be established taking into account the containment categories below and bearing in mind the specific circumstances of such operations.

| Specifications | Containment (EFB) categories | | |
	1	2	3
1) Viable micro-organisms should be contained in a system which physically separates the process from the environment (closed system):	yes	yes	yes
2) Exhaust gases from the closed system should be treated so as to:	minimise release	prevent release	prevent release
3) Sample collection, addition of materials to a closed system and transfer of viable micro-organisms to another closed system, should be performed so as to:	minimise release	prevent release	prevent release
4) Bulk culture fluids should not be removed from the closed system unless the viable micro-organisms to another closed system, should be performed so as	inactivated by validated means	inactivated by validated chemical or physical means	inactivated by validated chemical or physical means
5 Seals should be designed so as to:	minimise release	prevent release	prevent release
6) Closed systems should be located within a controlled area:	optional	optional	yes, and purpose-built
a) Biohazard signs should be posted	optional	yes	yes
b) Access should be restricted to nominated	optional	yes	yes, via airlock
c) Personnel should wear protective clothing	yes, work clothing	yes	a complete change
d) Decontamination and washing facilitites should be provided for personnel	yes	yes	yes
e) Personnel shoud shower before leaving	no	optional	yes

Table 6.4 Annex IV (from EC Council Directive of 23rd April, 1990 on the use of genetically modified micro-organisms).

| Specifications | Containment (EFB) categories | | |
	1	2	3
g) The controlled area should be adequately ventilated to minimise air contamination	optional	optional	yes
h) The controlled area should be maintained at an air pressure negative to atmosphere	no	optional	yes
i) Input air and extract air to the controlled area should be HEPA filtered	no	optional	yes
j) The controlled area should be designed to contain spillage of the entire contents of the closed system	optional	yes	yes
k) The controlled area should be sealable to permit fumigation	no	optional	yes
7) Effluent treatment before final discharge:	inactivated by validated means	inactivated by validated chemical or physical means	inactivated by validated chemical means

Table 6.4(Continued) Annex IV from EC Council Directive of 23rd April, 1990 on the use of genetically modified micro-organisms.

6.2.7 Additional provisions of the Directive

Other Articles follow which discuss public consultation procedures, confidentiality, information on safety planning and possible emergency measures to be taken in the event of an accident but these are beyond the scope of this chapter.

6.3 National Regulations and Contained Use of Genetically modified micro-organisms

Since the technology associated with genetically modifying organisms (genetic engineering) is relatively new, many national regulations were established before the EC-Directive described in section 6.2 was approved. As a result of this there is some disharmony between the various sets of regulations. Nevertheless, we must anticipate that now the EC-Directive is available there will be a progressive harmonisation of national regulations. Despite the differences that currently exist, many of the underpinning principles are common to all countries. We will use the approach used in the UK as an example of the application of a national approach to the regulation of the contained use of genetically modified micro-organisms. Of course, it is imperative to workers to conform to regulations and procedures which apply in their region/state.

6.3.1 Genetic Manipulation and Release in the UK

In the UK, controls on contained use of Genetically Modified Organisms (GMO's) are dealt with by the Genetic Manipulation Regulations 1989 which were brought into force through the multiple enabling powers conferred on the Secretary of State for Employment under provisions set out in the Health and Safety at Work etc. Act 1974. The 1989 Regulations completely replaced the earlier Health and Safety (Genetic

Manipulation) Regulations 1978 which have now been revoked. With regard to the release of (GMO's), the 1989 Regulations have been supplemented by provisions in the Environmental Protection Act 1990. Taken together, the various UK laws covering genetic manipulation, containment and release do satisfy the main provisions contained in the EC Directives 90/219/EEC (Directive on the contained use of genetically modified organisms) and 90/220/EEC (Directive on the deliberate release of genetically modified organisms into the environment) which relate to these topics.

For the purposes of this book, "Genetic manipulation" will be defined according to Health and Safety Executive of the United Kingdom Regulations published in 1989 which state : "Genetic manipulation means the propagation of combinations of hereditable material by the insertion of that material, prepared by whatever means outside a cell or organism, into a cell or organism in which it does not occur naturally, either:

a) directly; or

b) into a virus, microbial plasmid or other vector system which can then be incorporated in the cell or organism.

Also, according to these regulations, "Intentional introduction into the environment" means "the intentional introduction into the environment (that is outside provision for containment) of a live cell or organism which was produced or modified by genetic manipulation, *in vitro* by cell fusion or other *in vitro* technique, to form combinations of hereditable material which do not occur naturally in that cell or organism.

ACGM ACRE

Much of the discussion here will centre on these regulations but reference will be made also to proposed new regulations concerning genetically modified organisms published by the Health and Safety Commission in October 1991 in conjunction with the Department of the Environment, Ministry of Agriculture, Fisheries and Food, Scottish Office and the Welsh Office. Under the proposed new regulations, the Health and Safety Executive and the Department of the Environment will continue to take the lead in the formation and implementation of regulations regarding genetically modified organisms (GMO's) but the Advisory Committee on Genetic Modification (ACGM) and the Advisory Committee on Releases to the Environment (ACRE) will continue to provide expert advice.

Commencement of Genetic Manipulation (Modification)

Before an institution can start any work involving genetic modification as defined above, it must go through a number of administrative procedures.

1) Formation of a Genetic Modification Safety Committee

Such a committee, as well as being a statutory requirement does fulfil a number of very important roles not least of which is, of course, the establishing that the work being proposed and executed will be conducted in as safe a manner and situation as possible. Additionally, it is most important that the committee allays any worries that other employees in the Institute may have about the nature of the work taking place in the genetic modification laboratories and thus it is important that representatives of all personnel are represented on the committee.

Membership of the Genetic Modification Safety Committee

As mentioned above, it is important that the Safety Committee is properly constituted and representative. As well as containing representatives from the academic and technical staff working within the laboratory, it should also contain representatives of all staff who come in contact with the laboratory in any capacity eg secretaries, porters and cleaners. A representative of the management of the Department or Institute where the work is being carried out should be on the committee as well as the supervisory medical officer (although he/she would not normally be expected to attend regularly the meetings but would receive all minutes). The Biological Safety Officer of the Department or Institute is an important member of the committee and would normally act as committee chairperson.

Biological Safety Officer

role of BSO   It was a recommendation of the Williams Report (1976) that the post of Biological Safety Officer (BSO) is established for Genetic Modification Centres whose responsibility it would be to ensure that all local rules regarding genetic modification work are followed and that all personnel involved in the laboratory work have adequate training in appropriate microbiological practice. The BSO must be appropriately trained and experienced and should not be the Head of Department or Institute. The role of the BSO will vary in detail from one place to another but can best be described in summary as follows: The BSO:

- will ensure that all new members of staff (including research students) are competent to carry out the work expected of them in biological laboratories;

- will carry out periodic "safety audits" on laboratories and equipment;

- must be informed of any accidents or spillages in the laboratories;

- will liaise with the supervisory medical officer when necessary and with the Health and Safety Executive and other services who need to know of the genetic modification activity such as the local fire brigade;

- will ensure the safe storage of genetically modified organisms.

It is important that a deputy to the BSO be appointed so that absences can be covered and to bring an independent point of view to committee discussions when the BSO is personally involved with a specific experimental proposal.

2) Role of the Genetic Modification Safety Committee

The importance of local safety committees has been recognized ever since the publication of the first guidelines for genetic manipulation work. As mentioned earlier, as well as adjudicating on all aspects of risk associated with proposed work, the committee serves a most important function in public relations, in particular preventing unsubstantiated rumours from spreading round an Institute regarding the nature of the work taking place using genetic modification! Also, the committee must be satisfied that the laboratories in which genetic modification work takes place are suitable for the purpose and meet all the statutory and local requirements.

3) Starting Work on Genetic Modification

No person shall carry out genetic modification work unless he/she has notified the Health and Safety Executive in advance. The period of notification varies according to the type of work proposed. In the case of an activity involving intentional introduction into the environment, 90 days notice in advance is required. In all other cases, 30 days advance notice is required unless the Executive has agreed to a shorter period. The details required for initial notification are quite extensive and include:

- arrangements for physical containment;

- names of members of Genetic Modification Safety Committee;

- name of BSO and deputy BSO;

- name of supervisory medical officer;

- arrangements (if any) for health surveillance;

- comments made by the Genetic Modification Safety Committee on the local arrangements for risk assessment.

4) Risk Assessment

i) Hazard Groups of Organisms Used

The hazard group to which an organism is allocated dictates to a large extent the facilities which must be provided to carry out the basic microbiological work regardless of whether genetic modification is planned.

Group 1   An organism that is most unlikely to cause human disease.

Group 2   An organism that may cause human disease and which might be a hazard to laboratory workers but is unlikely to spread in the community. Laboratory exposure rarely produces infection and effective prophylaxis or effective treatment is usually available.

Group 3   An organism that may cause severe human disease and present a serious hazard to laboratory workers. It may present a risk of spread in the community but there is usually effective prophylaxis or treatment available.

Group 4   An organism that causes severe human disease and is a serious hazard to laboratory workers. It may present a high risk of spread in the community and there is usually no effective prophylaxis or treatment.

This classification was produced by the Advisory Committee on Dangerous Pathogens (see Chapter 5) and detailed information regarding the containment facilities required for working with organisms in the various hazard groups is available from them. Most genetic modification work involves organisms which fall into hazard groups 1 or 2 and may proceed in "normal" microbiology laboratories which do not have sophisticated containment facilities but have limited access.

ii) Risk Assessment in Genetic Modification

Guidelines were published in June 1988 by ACGM for the categorization of Genetic Manipulation Experiments (ACGM Note 7) and these guidelines form the basis of what is written below. These guidelines are enabled local genetic modification safety committees to assess the risk of individual proposals in a competent and consistent manner. This categorization scheme considers possible risks associated with genetic modification experiments under the headings of a) Access; b) Expression and c) Damage.

a) Access. This is a measure of the probability that a modified organism, or the DNA contained within it, will be able to enter the human body and survive there. It is important to note that this does not refer solely to enteric infections - other routes of infection must be considered. Typical relative values for the access factor are given below:

Known ability to colonize humans eg Wild-type *Escherichia coli*   1

Non-colonizing variants of a colonizing organism eg *E.coli* K12 10^{-3}

Disabled host/vector systems, whether laboratory constructed
or naturally occurring 10^{-6} to 10^{-9}

Genetically manipulated DNA in tissue culture cells introduced as DNA which does
not have the ability to infect or otherwise transfer to other cells 10^{-12}

b) Expression. This is a measure of the anticipated or known level of expression
of the inserted DNA. Typical examples are given below:

Deliberate in-frame insertion of expressible DNA downstream
of a promoter with the intention of maximising expression   1

Insertion of expressible DNA downstream of a promoter with
no attempts to maximise expression 10^{-3}

Insertion of expressible DNA into a site not specifically situated
to facilitate expression 10^{-6}

Non-expressible sequence 10^{-9}

c) Damage. This is a measure of the risk of a gene product resulting in ill-health of the worker exposed to a manipulated organism. Typical relative damage factors are given below:

Expression of a toxic substance or pathogenic determinant where
it is likely to have a significant deleterious biological effect   1

Expression of a biologically active substance which might have a
deleterious effect if it were delivered to a target tissue 10^{-3}

Expression of a biologically active molecule which is unlikely to have a deleterious
effect, for example, when it could not approach the normal body level 10^{-6}

Use of a gene sequence where any biological effect is considered
to be highly unlikely 10^{-9}

No forseeable biological effect 10^{-12}

d) Risk Assessment in relation to the physical containment required.

The overall risk assessment is the product of the figures assessed for Access, Expression and Damage. It is this overall assessment which determines the necessary level of containment:

Risk Assessment	Containment Level
10^{-15} or lower	1
10^{-12} to 10^{-14}	2
10^{-9} to 10^{-11}	3
10^{-6} to 10^{-9}	case by case
10^{-5} or greater	4

As mentioned above, Containment levels 1 and 2 require no special facilities over and above those found in good microbiology laboratories, but Containment levels 3 and 4 do require special facilities and the opinions of the Health and Safety Executive (HSE) must be sought in these situations.

The local Genetic Modification Safety Committee must satisfy itself regarding the risk assessment as judged by the person responsible for the project and the data on which it is based. It must also be satisfied with regard to the level of containment of the project. The chairman of the local safety committee is responsible for the annual return to HSE of all genetic modification activities.

e) Codes of Practice

It is advisable that all centres of genetic modification publish codes of practice outlining local safety rules and guidance for personnel working in, or coming in contact with, the laboratories where such work is carried out. These rules should encompass behaviour in the laboratory and disinfection policy etc. as well as advice to cleaners and porters who may have cause to enter the laboratory. Such codes of practice should only be issued with the approval and consent of the local genetic modification safety committee.

f) Proposed New Rules

The Department of the Environment together with the HSE have proposed new rules (Genetically Modified Organisms [Environmental Protection] Regulations) which impose specific requirements based on the general principles set out in the Environmental Protection Act 1990. These regulations have all the main features of the existing rules under the Genetic Manipulation Regulations (1989) but in addition contain some significant new additions including:

1) an expansion of the classification of genetically manipulated micro-organisms to take into account environmental implications.

2) the need to notify the Secretary of State of the Environment as well as the HSE in certain cases. Fees will be charged for processing consent applications. Under the proposed regulations, these fees may be substantial (eg currently ranging from £ 2 000 to £ 4 000 per consent).

6.4 EC Directive on the deliberate release of genetically modified organisms (90/220/EEC)

6.4.1 Parts of the Directive

This directive consists of four parts:

Part A - General provisions;

Part B - Research and Development (R & D) and other introductions into the environment than placing on the market;

Part C - Placing of products on market ;

Part D - Final provisions.

6.4.2 Part A: General provisions

Purpose of the directive

The purpose of this directive is laid down in Article 1:

"to approximate the laws, regulations and administrative provisions of the Member State and to protect human health and the environment when carrying out a deliberate release or placing on the market of genetically modified organisms".

In order to gain a better understanding of the purpose and background of this directive, the considerations in the preamble should also be studied.

In addition to the purpose of this directive Article 4 emphasises in general terms the obligations of member states in accomplishing this purpose.

Scope of the directive-definitions

Repeating what was explained earlier: the scope of every regulation depends on the definitions and exemptions.

This directive covers the deliberate release of genetically modified organisms. In the following paragraphs the definitions of the directive are explained. These definitions are summed up in Article 2.

Organism

Organism is defined in this directive as:

"any biological entity, capable of replication or of transferring genetic material".

Since the term 'biological entity' is open for multiple interpretation, an explanation is given in the statements for inclusion in the Council's minutes:

"This definition covers: micro-organisms, including viruses and viroids; plants and animals; including ova, seeds, pollen, cell cultures and tissue cultures from plants and animals".

Hereafter, the term genetically modified organisms is abbreviated to GMO.

Modified organism

The definition of a genetically modified organism is analogous to the definition of a genetically modified micro-organisms, provided that the term 'micro-organism' is replaced by the term 'organism'.

Deliberate release

Deliberate release is defined in paragraph 3 of Article 2 as:

"any intentional introduction into the environment of a GMO or a combination of GMOs without provisions for containment such as physical barriers or a combination of physical barriers together with chemical and/or biological barriers used to limit their contact with the general population and the environment".

A further clarification is given in the Council's Statements:

"the introduction by whatever means, directly or indirectly, by using, storing, disposing, or making available to a third party".

By using the terms "without provisions for containment such as..." as a cross reference to the directive for contained use, a complementary system is achieved.

In other words: every activity that is not a contained use is regarded as a deliberate release.

Exemptions

The exemptions of the scope of this directive are laid down in Article 3:

"This directive shall not apply to organisms obtained through the techniques of genetic modification listed in Annex Ib". These include:

- mutagenesis;

- cell fusion of plant cells when the plant can also be produced by traditional methods.

The background of these exemptions is that these specific applications have been used in a number of applications and have a long safety record.

6.4.3 System

The system of this directive is based on two notions:

- the release of a GMO into the environment can have adverse effects on the environment which may be irreversible;

- GMOs, as well as other organisms, are not stopped by national frontiers.

These two notions led to the choice of a system whereby:

- every introduction of a GMO into the environment is subject to an authorization by the competent authority of the country where the introduction takes place;

- before an authorization is given, the competent authority consults the other Member States of the Community.

In addition to this, a distinction between Research and Development (R & D) and placing on the market is made.

R & D and placing on the market

In this directive, a distinction is made between:

Research and Development (R & D) and introductions into the environment other than piacing on the market (part B of the directive);

Placing on the market (part C of the directive).

The result of this distinction and the system of the directive is that placing on the market involves a system of international consultation whereby the competent authority can not take a decision without the agreement of the other Member States.

All other introductions into the environment (Part B, which basically consists of R & D introductions) are subject to an authorization of the competent authority which may give its decision without the approval of other Member States, though be it that these introductions are also notified to the other Member States who may give comments.

The reason for this distinction, is found in the numbers and the spread of a GMO connected with placing on the market. When a product containing GMOs or consisting of a GMO is placed on the market, it will be spread all over Europe in, possibly, vast numbers under uncontrolled circumstances. Whereas R & D introductions are normally small scale introductions of a limited number of GMOs and under controlled circumstances.

6.4.4 Part B: Research and Development (R & D) and introductions into the environment other than placing on the market

The basis of part B of this directive is laid down in the combination of the Articles 5, paragraph 1 and Article 6, paragraph 4.

Article 5, paragraph 1 states:

"Any person before undertaking a deliberate release of a GMO for the purpose of research and development or for any other purpose than placing on the market, must submit a notification to the competent authority of the Member State within whose territory the release is to take place".

Article 6, paragraph 4, states:

"The notifier may proceed with the release only when he has received the written consent of the competent authority, in conformity with any conditions required in this consent".

The notification

Article 5, paragraph 2, gives the requirements of a notification under part B of the directive:

"The notification shall include the information specified in Annex II".

Annex II is an indicative list of points of information set out under 5 headings:

  I   General information

  II   Information related to the GMO

  III Information relating to the conditions of release and the receiving environment

  IV   Information relating to the GMO and the environment

  V   Information on monitoring, control, waste treatment and emergency response plans

This Annex II is based on the OECD report of 1986. It is essential to realise that this Annex contains an indicative list, and that not all the points included will apply to every case.

Authorisation

The authorisation procedure is laid down in article 6.

Paragraph 1: "On receipt and after acknowledgement of the notification the competent authority shall examine the conformity of the notification with the requirements of this directive".

Paragraph 2: "The competent authority, having considered where appropriate, any comments by other Member States, shall respond in writing to the notifier within 90 days by indicating either:

* that the release may proceed;

* that the release does not fulfil the conditions of this directive and the notification is therefore rejected.

For calculating the waiting period of 90 days, the period needed for the notifier to supply further information and the period in which a public inquiry is carried out, shall not be taken into account.

The third paragraph of Article 6 gives the steps to be taken when new information becomes available with regard to the risk of the product. In that case the notifier shall revise the information, inform the competent authority and take the necessary measures to protect human health and the environment.

International consultation

Within 30 days after the receipt of a notification, the competent authority shall send to the Commission a summary of the notification. The Commission shall immediately forward these summaries to the other Member States which may, within 30 days, present observations. It should be stressed here that these observations are not binding to the original competent authority.

6.4.5 Part C: Placing on the market products containing genetically modified organisms

General provisions

Part C of this directive stars with Article 10, which gives in paragraph 1, a set of general conditions before any product can be placed on the market.

These conditions are that:

* consent has been given under part B of the directive, meaning that no GMO can be placed on the market without a proper R & D stage;

* the product should comply with this directive and relevant product legislation.

Paragraph 2 of Article 10 indicates that the procedure for placing a product on the market shall not apply to products covered by Community legislation which includes a specific environmental risk assessment similar to that provided in this directive. The background of this provision is that:

* it is desirable to have only one procedure for placing the products on the market;

* product legislation already contains procedures for placing on the market.

System

The same system of part B is found in part C.

Article 11, paragraph 1:

"before a GMO or a combination of GMOs are placed on the market as such or in a product, the manufacturer or the importer to the Community shall submit a notification to the competent authority of the Member State where they are placed on the market for the first time".

Article 11, paragraph 5:

"the notifier may only proceed when he has received a written consent".

The notification

The first paragraph of Article 11 says that the notification shall include the information of Annex II, information obtained from R & D releases and specific product information laid down in annex III (use, labelling, packaging etc).

The final paragraph of Article 11 gives the steps to be taken when new information becomes available with regard to the risk of the product, analogous to Article 6.

Authorisation

The authorisation procedure of placing on the market is in fact a two step procedure. The first step is given by Article 12:

Paragraph 1: "On receipt and after acknowledgement of the notification the competent authority shall examine the conformity of the notification with the requirements of this directive".

Paragraph 2: "The competent authority shall respond within 90 days by either:

• forwarding the dossier to the Commission with a favourable opinion;

• informing the notifier that the release does not fulfil the conditions of this directive and the notification is therefore rejected.

For calculating the waiting period of 90 days the period needed for the notifier to supply further information shall not be taken into account.

International consultation

The second step of the procedure is laid down in Article 13:

"The Commission shall immediately forward the dossier to the other Member States which may, within 60 days, present observations which are received from the other Member States".

When an objection is received and the competent authorities concerned cannot reach an agreement within these 60 days, the commission shall take a decision in accordance to a specific procedure.

Placing on the market: Community wide

One of the key articles of this part C is Article 15:

"A Member State may not restrict or impede, on grounds relating to the notification and written consent of a release under this directive, the placing on the market of product containing or consisting of GMOs which comply with the requirements of this directive".

This means that when a product has received a consent after the procedure of part C, no Member State may restrict the placing on the market on grounds of protecting human health or the environment.

When a Member State has justifiable reasons (e.g. new information) that such a product constitutes a risk, it may provisionally restrict the product, after which the Commission shall take a decision in accordance to a specific procedure (Article 16).

The commission shall publish a list of products which received consent under this directive (Article 18).

6.4.6 Part D: Final provisions

Confidentiality

Article 19:

"The Commission and the competent authorities shall not divulge to third parties any confidential information notified or exchanged under this directive and shall protect the intellectual property rights relating to the data received".

The notifier indicates what information he wants to be kept confidential, though be it that certain information cannot be kept confidential, like the name and address of the notifier and a description of the GMO, methods for monitoring and the evaluation of foreseeable effects.

It is the competent authority which decides, after consultation with the notifier, which information shall be kept confidential.

Commission procedure

The specific procedure mentioned before is explained in Article 21, which in general terms says that the Commission will be assisted by a Committee which votes by qualified majority. If measures envisaged by the Commission are not in accordance with the opinion of the committee, it will be submitted to the Council.

6.5 Risk analysis and the deliberate release of deliberate genetically modified organisms

As we described in section 6.3 for the contained use of genetically modified micro-organisms, there is some disharmony between the systems used in various Member States. Nevertheless EC-Directive 90/220/EEC will do much to harmonise procedures and standards Here we will consider the features of the risk analysis that should be conducted prior to release of a transgenic organism.

In carrying out a risk analysis for the release of a genetically modified organism the following factors must be included:

- Characteristics of the host:

    description of host;

    potential harmfulness of the transgenic organism;

    growth and survival;

    genetic stability.

- Characteristic of the genetic inserts:

    donor DNA;

    protein eroded by the trangene.

- Characteristic of the transgenic organism;

    method of construction.

- Description of the release;

    aim;

    location;

    emergency measures.

Scientific Procedure using animals

Appendix 7.1

Scientific Procedure using animals

7.1 Introduction

Much public debate centres on the ethical issues concerning the use of animals in scientific procedures. Despite such concern, the use of animals is, however, likely to continue for the foreseeable future. The use of animals has however been strictly controlled for many years through wide ranging legislation. Initially these nationally-based regulations and recommendations showed quite marked differences. The approval of EC-Directive 86/609/EEC is encouraging elimination of disparities between Member States regarding the protection of animals used for experimental and other scientific purposes. Since EC-Directive 86/609/EEC provides the common basis of the regulations governing the use of animals for experimental purposes, we will largely confine discussion to this Directive.

7.2 Activities covered

Directive 86/609/EEC applies to the use of animals in experiments for any of the following purposes:

- the development, manufacture, quality, effectiveness and safety testing of drugs, foodstuffs and other substances or products:

  to avoid, prevent, diagnose or treat disease, ill health or other abnormalities or their effects in man, animals or plants;

  for the assessment, detection, regulation or modification of physiological conditions in man, animals or plants.

- the protection of the natural environment in the interests of the health or welfare of man or animals.

7.3 Overview of EC Directive 86/609/EEC

The title of the Directive is 'Council Directive of 24th November 1986 on the approximation of laws, regulations and administrative provisions of the Member States regarding the protection of animals used for experimental and other scientific purposes.' It is published in the Official Journal No L385/1.

The Directive is quite short consisting of 27 articles and is reproduced in full as an appendix to this chapter. The main topics covered are:

- general statements regarding accommodation, anaesthesia, euthanasia, re-use of animals, release of animals, breeding;

- establishment of one (or more) authorities within each Member State with responsibility for verifying whether the requirements are met (see Section 7.3);

- Member States have responsibility of providing the European Committee with all relevant information regarding quantities of animals used and of the results of experiments that have been performed. Confidential information is, however, treated as such;

- Member States are obliged to recognise the validity of the data generated in other Member States;

- the search for alternatives, avoiding or reducing the use of animals; should be encouraged;

- Member States, although obliged to follow the conditions specified in the Directive, are specifically enabled by the Directive to impose stricter measures for the protection of animals if they so desire.

7.4 Local inspection

establishment of an inspectorate

A key point of the Directive is the need for each Member State to establish one (or more) authorities to monitor the use of animals for scientific purposes. In the Netherlands, this is done, at a national level, by the Veterinary Inspectorate. Each Institution or company using animals has to establish a committee to control this use of animals.

In the UK, all animal experiments are regulated by the Animals (Scientific Procedures) Act 1986 which replaced and repealed the Cruelty to Animals Act 1876. This Act brought UK regulations in line with the requirements of EC-Directive 86/69/EEC. Thus the main method of control is through the issuing and use of licences. Two separate licences are required before animal experiments can be performed. Firstly a personal licence is required by the experimenter through application to the Home Office. Secondly, a project licence is required to authorise work on a particular project or programme. The 1986 Act sets out the specific areas towards which projects must be directed if they are to be granted a licence. These include study of diseases, study of physiological conditions, protection of the environment for man or animals, research in biological or behaviourial sciences, education and training (eg surgical techniques for medical or veterinary students), forensic testing and breeding animals for experimental or other scientific uses. The 1986 Act also specifies the nature of the living entities covered; all vertebrates other than Man, beginning in the cases of mammals, birds or reptiles, after half-way through their gestational or incubation period. For other vertebrates, the Act applies once the animal becomes capable of independent feeding.

The 1986 Act also specifies the types of procedures that are regulated; this includes any experimental or scientific procedure which may cause a protected animal pain, distress or lasting harm. (The phrase 'protected animal' means those that are covered by the Act). Moreover, it is this Act that would apply in cases of genetic manipulation of vertebrate animals other than man. Thus, production of transgenic vertebrates would be an offence unless carried out under the licensing system and other control measures of the 1986 Act.

The 1986 Act is enforced in practice by an Inspectorate and the Secretary of State is advised on matters concerned with the Act by an advisory committee, the Animal Procedures Committee. Home Office Guidelines are issued concerning operations of the 1986 Act as well as the general and specific conditions for obtaining licences. It is fair to say that UK law satisfies the EC standards (86/609/EEC) in nearly all respects.

7.5 EC - Guidelines for animal welfare

The conditions specified within EC-Directive 86/609/EEC are, of course, obligatory for all Member States. However, a set of guidelines of the accommodation and care of animals has been added to the Directive in the form of an annex. These guidelines contain detailed information on the sizes of facilities needed to accommodate animals, temperature of housing facilities, quarantine periods etc. The guidelines are not obligatory. The more lenient nature of this annex is well illustrated by the statement that existing housing should not be replaced before it is worn out!

In the United States a similar set of guidelines are published in a 'Guide for the Care and Use of Laboratory Animals. This was first issued in 1963 and has been periodically updated (for example "Guide for the care and Use of Laboratory Animals, US Department of Health and Human Services 1985, NIH Publication No 86-23 revised 1985) The American document contains much more detail than the European Directive.

Appendix 7.1

OJ. No L 358/1
(18.12.86)

Council Directive of 24 November 1986

on the approximation of laws, regulations and administrative provisions of the Member States regarding the protection of animals used for experimental and other scientific purposes

(86/609/EEC)

Having regard to the Treaty establishing the European Economic Community, and in particular Article 100 thereof.

Having regard to the proposal from the Commission (¹),

Having regard to the opinion of the European Parliament (²),

Having regard to the opinion of the Economic and Social Committee (³).

Whereas there exist between the national laws at presetn in force for the protection of animals used for certain experimental purposes disparities which may affect the functioning of the common market;

Whereas, in order to eliminate these disparities, the laws of the Member States should be harmonized; whereas such harmonisation should ensure that the number of animals used for experimental or other scientific purposes is reduced to a minimum, that such animals are adequately cared for, that no pain, suffering, distress or lasting harm are inflicted unnecessarily and ensure that, where unavoidable, these shall be kept to the minimum;

Whereas, in particular, unnecessary duplication of experiments should be avoided.

Article 1

The aim of this Directive is to ensure that where animals are used for experimental or other scientific purposes the provisions laid down by law, regulations or administrative provisions in the Member States for their protection are approximated so as to avoid affecting the establishment and functioning of the common market, in particular by distortions of competition or barriers to trade.

Article 2

For the purposes of this Directive the following definitions shall apply:

(a) 'animal' unless otherwise qualified, means any live non-human vertebrate, including free-living

larval and/or reproducing larval forms, but excluding foetal or embryonic forms;

(b) 'experimental animals' means animals used or to be used in experiments;

(c) 'bred animals' means animals specially bred for use in experiments in facilities approved by, or registered with, the authority;

(d) 'experiment' means any use of an animal for experimental or other scientific purposes which may cause it pain, suffering, distress or lasting harm, including any course of action intended, or liable, to result in the birth of an animal in any such condition, but excluding the least painful methods accepted in modern practice (i.e. 'humane' methods) of killing or marking an animal; an experiment starts when an animal is first prepared for use and ends when no further observations are to be made for that experiment; the elimination of pain, suffering, distress or lasting harm by the successful use of anaesthesia or analgesia or other methods does not place the use of an animal outside the scope of this definition. Non experimental, agricultural or clinical veterinary practices are excluded.

(e) 'authority' means the authority or authorities designated by each Member State as being responsible for supervising the experiments within the meaning of this Directive;

(f) 'competent person' means any person who is considered by a Member State to be competent to perform the relevant function described in this Directive;

(g) establishment' means any installation, building, group of buildings or other premises and may include a place which is not wholly enclosed or covered and mobile facilities;

(h) 'breeding establishment' means any establishment where animals are bred with a view to their use in experiments;

(i) 'supplying establishment' means any establishment, other than a breeding establishment, from which animals are supplied with a view to their use in experiments;

(1) OJ No C 351, 31. 12.1985, p.16.
(2) OJ No C 255, 13. 10. 1986, p. 250.
(3) OJ No C 207, 18.8. 1986, p. 3.

(j)'user establishment' means any establishment, where animals are used for experiments;

(k)'properly anaesthetized' means deprived of sensation by methods of anaesthesia (whether local or general) as effective as those used in good veterinary practice

(l) 'humane method of killing' means the killing of an animal with a minimum of physical and mental suffering depending on the species.

Article 3

This Directive applies to the use of animals in experiments which are undertaken for one of the following purposes:

(a) the development, manufacture, quality, effectiveness and safety testing of drugs, foodstuffs and other substances or products:

(i) for the avoidance, prevention, diagnosis or treatment of disease, ill-health or other abnormality or their effects in man, animals or plants;

(ii) for the assessment, detection, regulation or modification of physiological conditions in man, animals or plants;

(b) the protection of the natural environment in the interests of the health or welfare of man or animal.

Article 4

Each Member State shall ensure that experiments using animals considered as endangered under Appendix 1 of the Convention on International Trade in Endangered Species of Fauna and Flora and Annex C.I. of Regulation (EEC) No 3626/82 (^{1}) are prohibited unless they are in conformity with the above Regulation and the objects of the experiment are:-

research aimed at preservation of the species in question, or -

essential biomedical purposes where the species in question exceptionally proves to be the only one suitable for those purposes.

Article 5

Member States shall ensure that, as far as the general care and accommodation of animals is concerned:

(a) all experimental animals shall be provided with housing, an environmental, at least some freedom of movement, food, water and care which are appropriate to their health and well-being;

(b) any restriction on the extent to which an experimental animal can satisfy its physiological and ethological needs shall be limited to the absolute minimum;

(1) OJ No L 384, 31.12.1982, p.1.

(c) the environmental conditions in which experimental animals are bred, kept or used must be checked daily;

(d) the well-being and state of health of experimental animals shall be observed by a competent person to prevent pain or avoidable suffering, distress or lasting harm;

(e) arrangements are made to ensure that any defect or suffering discovered is eliminated as quickly as possible.

For the implementation of the provisions of paragraphs (a) and (b). Member States shall pay regard to the guidelines set out in Annex II.

Article 6

1. Each Member State shall designate the authority or authorities responsible for verifying that the provisions of this Directive are properly carried out.

2. In the framework of the implementation of this Directive, Member States shall adopt the necessary measures in order that the designated authority mentioned in paragraph 1 above may have the advice of experts competent for the matters in question.

Article 7

1. Experiments shall be performed solely by competent authorised persons, or under the direct responsibility of such a person, or if the experimental or other scientific project concerned is authorised in accordance with the provisions of national legislation.

2. an experiment shall not be performed if another scientifically satisfactory method of obtaining the result sought, not entailing the use of an animal, is reasonably and practicably available.

3. When an experiment has to be performed, the choice of species shall be carefully considered and, where necessary explained to the authority. In a choice between experiments, those which use the minimum number of animals, involve animals with the lowest degree of neurophysiological sensitivity, cause the least pain, suffering, distress or lasting harm and which are most likely to provide satisfactory results shall be selected.

Experiments on animals taken from the wild may not be carried out unless experiments on other animals would not suffice for the aims of the experiment.

4. All experiments shall be designed to avoid distress and unnecessary pain and suffering to the experimental animals. They shall be subject to the provisions laid down in Article The measures set out in Article 9 shall be taken in all cases.

Article 8

1. All experiments shall be carried out under general or local anaesthesia.

2. Paragraph 1 above does not apply when:

(a) anaesthesia is judged to be more traumatic to the animal than the experiment itself;

(b) anaesthesia is incompatible with the object of the experiment. In such cases appropriate legislative and/or administrative measures shall be taken to ensure that no such experiment is carried out unnecessarily.

Anaesthesia should be used in the case of serious injuries which may cause severe pain.

3. If anaesthesia is not possible, analgesics or other appropriate methods should be used in order to ensure as far as possible that pain suffering, distress or harm are limited and that in any event the animal is not subject to severe pain, distress or suffering.

4. Provided such action is compatible with the object of the experiment, an anaesthetized animal, which suffers considerable pain once anaesthesia has worn off, shall be treated in good time with pain-relieving means or, if this is not possible, shall be immediately killed by a humane method.

Article 9

1. At the end of any experiment, it shall be decided whether the animal shall be kept alive or killed by a humane method, subject to the condition that it shall not be kept alive if, even though it has been restored to normal health in all other respects, it is likely to remain in lasting pain or distress.

2. The decisions referred to in paragraph 1 shall be taken by a competent person, preferably a veterinarian.

3. Where at the end of an experiment:

(a) an animal is to be kept alive, it shall receive the care appropriate to its state of health, be placed under the supervision of a veterinarian or other competent person and shall be kept under conditions confirming to the requirements of Article 5. The conditions laid down in this subparagraph may, however, be waived where, in the opinion of a veterinarian, the animal would not suffer as a consequence of such exemption;

(b) an animal is not to be kept alive or cannot benefit form the provisions of Article 5 concerning its well-being, it shall be killed by a humane method as soon as possible.

Article 10

Member States shall ensure that any re-use of animals in experiments shall be compatible with the provisions of this Directive.

In particular, an animal shall not be used more than once in experiments entailing severe pain, distress or equivalent suffering.

Article 11

Notwithstanding the other provisions of this Directive, where it is necessary for the legitimate purposes of the experiment, the authority may allow the animal concerned to be set free, provided that it is satisfied that the maximum possible care has been taken to safeguard the animals's well-being, as long as its state of health allows this to be done and there is no danger for public health and the environment.

Article 12

1. Member States shall establish procedures whereby experiments themselves or the details of persons conducting such experiments shall be notified in advance to the authority.

2. Where it is planned to subject an animal to an experiment in which it will, or may, experience severe pain which is likely to be prolonged, that experiment must be specifically declared and justified to, or specifically authorised by, the authority. The authority shall take appropriate judicial or administrative action if it is not satisfied that the experiment is of sufficient importance for meeting the essential needs of man or animal.

Article 13

1. On the basis of requests for authorisation and notifications received, and on the basis of the reports made, the authority in each Member State shall collect, and as far as possible periodically make publicly available, the statistical information on the use of animals in experiments in respect of:

(a) the number and kinds of animals used in experiments;

(b) the number of animals, in selected categories, used in the experiments referred to in Article 3;

(c) the number of animals, in selected categories, used in experiments required by legislation.

2. Member States shall take all necessary steps to ensure that the confidentiality of commercially sensitive information communicated pursuant to this Directive is protected.

Article 14

Persons who carry out experiments or take part in them and persons who take care of animals used for experiments, including duties of a supervisory nature, shall have appropriate education and training.

In particular, persons carrying out or supervising the conduct of experiments shall have received instruction in a scientific discipline relevant to the experimental work being undertaken and be capable of handling and taking care of laboratory animals; they shall also have satisfied the authority that they have attained a level of training sufficient for carrying out their tasks.

Article 15

Breeding and supplying establishments shall be approved by or registered with, the authority and comply with the requirements of Articles 5 and 14 unless an exemption is granted under Article 19 (4) or Article 21. A supplying establishment shall obtain animals only from a breeding or other supplying establishment unless the animal has been lawfully imported and is not a feral or stray animal. General or special exemption from this last provision may be granted to a supplying establishment under arrangements determined by the authority.

Article 16

The approval or the registration provided for in Article 15 shall specify the competent person responsible for the establishment entrusted with the task of administering, or arranging for the administration of, appropriate care to the animals bred or kept in the establishment and of ensuring compliance with the requirements of Articles 5 and 14.

Article 17

1. Breeding and supply establishments shall record the number and the species of animals sold or supplied, the dates on which they are sold or supplied, the name and the address of the recipient and the number and species of animals dying while in the breeding or supplying establishment in question.

2. Each authority shall prescribe the records which are to be kept and made available to it by the person responsible for the establishments mentioned in paragraph 1; such records shall be kept for a minimum of three years from the date of the last entry and shall undergo periodic inspection by officers of the authority.

Article 18

1. Each dog, cat or non-human primate in any breeding, supplying or user establishment shall, before it is weaned, be provided with an individual identification mark in the least painful manner possible except in the cases referred to in paragraph 3.

2. Where an unmarked dog, cat or non-human primate is taken into an establishment for the first time after it has been weaned it shall be marked as soon as possible.

3. Where a dog, cat or non-human primate is transferred from one establishment as referred to in paragraph 1 to another before it is weaned, and it is not practicable to mark it beforehand, a full documentary record, specifying in particular its mother, must be maintained by the receiving establishment until it can be so marked.

4. Particulars of the identity and origin of each dog, cat or non-human primate shall be entered in the record of each establishment.

Article 19

1. User establishments shall be registered with, or approved by, the authority. Arrangements shall be made for user establishments to have installations and equipment suited to the species of animals used and the performance of the experiments conducted there; their design, construction and method of functioning shall be such as to ensure that the experiments are performed as effectively as possible, with the object of obtaining consistent results with the minimum number of animals and the minimum degree of pain, suffering, distress or lasting harm.

2. In each user establishment:

(a) the person or persons who are administratively responsible for the care of the animals and the functioning of the equipment shall be identified;

(b) sufficient trained staff shall be provided;

(c) adequate arrangements shall be made for the provision of veterinary advice and treatment;

(d) a veterinarian or other competent person should be charged with advisory duties in relation to the well being of the animals.

3. Experiments may, where authorised by the authority, be conducted outside user establishments.

4. In user establishments, only animals from breeding or supplying establishments shall be used unless a general or special exemption has been obtained under arrangements determined by the authority. Bred animals shall be used whenever possible. Stray animals of domestic species shall not be used in experiments. A general exception made under the conditions of this paragraph may not extend to stray dogs and cats

5. User establishments shall keep records of all animals used and produce them whenever required to do so by the authority. In particular, these records shall show the number and species of all animals acquired, from whom they were acquired and the date of their arrival. Such records shall be kept for a minimum of three years and shall be submitted to the authority which asks for them User establishments shall be subject to periodic inspection by representatives of authority.

Article 20

When user establishments breed animals for use in experiments on their own premises, only one registration or approval is needed for the purposes of Article 15 and 19 However, the establishments shall comply with the relevant provisions of this Directive concerning breeding and user establishments.

Article 21

Animals belonging to the species listed in Annex I which are to be used in experiments shall be bred animals unless a general or special exemption has been obtained under arrangements determined by the authority.

Article 22

1. In order to avoid unnecessary duplication of experiments for the purposes of satisfying national or Community health and safety legislation, Member States shall as far as possible recognise the validity of data generated by experiments carried out in the territory of another Member State unless further testing is necessary in order to protect public health and safety.

2. To that end, Member States shall, where practicable and without prejudice to the requirements of existing Community Directives, furnish information to the Commission on their legislation and administrative practice relating to animal experiments, including requirements to be satisfied prior to the marketing of products; they shall also supply factual information on experiments carried out in their territory and on authorisations or any other administrative particulars pertaining to these experiments.

3. The Commission shall establish a permanent consultative committee within which the Member States would be represented, which will assist the Commission in organizing the exchange of appropriate information, while respecting the requirements of confidentiality, and which will also assist the Commission in the other questions raised by the application of this Directive.

Article 23

1. The Commission and Member States should encourage research into the development and validation of alternative techniques which could provide the same level of information as that obtained in experiments using animals but which involve fewer animals or which entail less painful procedures, and shall take such other steps as they consider appropriate to encourage research in this field. The Commission and Member States shall monitor trends in experimental methods.

2. The Commission shall report before the end of 1987 on the possibility of modifying tests and guidelines laid down in existing Community legislation taking into account the objectives referred to in paragraph 1.

Article 24

This Directive shall not restrict the right of the Member States to apply or adopt stricter measures for the protection of animals used in experiments or for the control and restriction of the use of animals for experiments. In particular, Member States may require a prior authorisation for experiments or programmes of work notified in accordance with the provisions of Article 12(1).

Article 25

1. Member States shall take the measures necessary to comply with this Directive by 24 November 1989. They shall forthwith inform the Commission thereof.

2. Member States shall communicate to the Commission the provisions of national law which they adopt in the field covered by this Directive.

Article 26

At regular intervals not exceeding three years, and for the first time five years following notification of this Directive, Member States shall inform the Commission of the measures taken in this area and provide a suitable summary of the information collected under the provisions of Article 13. The Commission shall prepare a report for the Council and the European Parliament0

Article 27

This Directive is addressed to the Member States.

Done at Brussels, 24 November 1986.

For the Council

The President

ANNEX I

List of experimental animals covered by the provisions of article 21

- Mouse - *Mus musculus*

- Rat - *Rattus norvegicus*

- Guinea Pig - *Cavia procellus*

- Golden Hamster - *Mesocricetus autratus*

- Rabbit - *Oryctolagus cuniculus*

- Non-human primates

- Dog - *Canis familiaris*

- Cat - *Felis catus*

- Quail - *Coturnix coturnix*

ANNEX II

Guidelines for accommodation and Care of Animals

(Article 5 of the Directive)

Introduction

1) The Council of the European Economic Community has decided that the aim of the Directive is to harmonise the laws of the Member States for the protection of animals used for experimental and other scientific purposes in order to eliminate disparities which at present may affect the functioning of the common market. 'Harmonisation should ensure that such animals are adequately cared for, that no pain, suffering, distress or lasting harm are inflicted unnecessarily and that where unavoidable the latter shall be kept to the minimum.

2) It is true that some experiments are conducted under field conditions on free-living, self-supporting, wild animals, but such experiments are relatively few in number. The great majority of animals used in experiments must for practical reasons be kept under some sort of physical control in facilities ranging from outdoor corrals to cages for small animals in a laboratory animal house. This is a situation where there are highly conflicting interests. On the one hand, the animal whose needs in respect of movement, social relations and other manifestations of life must be restricted, on the other hand, the experimenter and his assistants who demand full control of the animal and its environment. In this confrontation of interests the animal may sometimes be given secondary consideration.

3) Therefore, the Directive provides in Article 5 that: 'as far as the general care and accommodation of animals is concerned:

   a) all experimental animals shall be provided with housing, an environment, as least some freedom of movement, food, water and care which are appropriate to their health and well-being;

   b) any restriction on the extent to which an experimental animal can satisfy its physiological and ethological needs shall be limited to the absolute minimum'.

4) This Annex draws up certain guidelines based on present knowledge and practice for the accommodation and care of animals. It explains and supplements the basic principles adopted in Article 5. The object is thus to help authorities, institutions and individuals in their pursuit of the aims of the Directive in this matter.

5) Care is a word which, when used in connection with animals intended for or in actual use in experiments covers all aspects of the relationship between animals and man. Its substance is the sum of material and non-material resources mobilised by man to obtain and maintain an animal in a physical and mental state where it suffers least and performs best in experiments. It starts from the moment the animal is destined to be used in experiments and continues until it is killed by a humane method or otherwise disposed of by the establishment in accordance with Article 9 of the Directive after the close of the experiment.

6) This Annex aims to give advice about the design of appropriate animal quarters. There are, however, several methods of breeding and keeping laboratory animals that differ chiefly in the degree of control of the microbiological environment. It has to be borne in mind that the staff concerned will sometimes have to judge from the character and condition of the animals where the recommended standards of space may not be sufficient, as with especially aggressive animals. In applying the guidelines described in this Annex the requirements of each of these situations should be taken into account. Furthermore, it is necessary to make clear the status of these guidelines. Unlike the provisions of the Directive itself, they are not mandatory; they are recommendations to be used with discretion, designed as guidance to the practices, and standards which all concerned should conscientiously strive to achieve. It is for this reason that the term 'should' has had to be used throughout the text even where 'must' might seem to be the more appropriate word. For example, it is self-evident that food and water *must* be provided (see 3.7.2 and 3.8).

7) Finally, for practical and financial reasons, existing animal quarters equipment should not need to be replaced before it is worn out and has otherwise become useless. Pending replacement with equipment conforming with the present guidelines, these should as far as practicable be complied with by adjusting the numbers and sizes of animals placed in existing cages and pens.

DEFINITIONS

In this Annex, in addition to the definitions contained in Article 2 of the Directive:

a) 'holding rooms' means rooms where animals are normally housed, either for breeding and stocking or during the conduct of an experiment;

b) 'cage' means a permanently fixed or movable container that is closed by solid walls and, at least on one side, by bars or meshed wire or, where appropriate nets and in which one or more animals are kept or transported; depending on the stocking density and the size of the container, the freedom of movement of the animals is relatively restricted;

c) 'pen' means an area enclosed, for example, by walls, bars or meshed wire in which one or more animals are kept; depending on the size of the enclosure and the stocking density the freedom of movement of the animals is usually less restricted than in a cage;

d) 'run' means an area closed, for example, by fences, walls, bars or meshed wire and frequently situated outside permanently fixed buildings in which animals kept in cages or pens can move freely during certain periods of time in accordance with their ethological and physiological needs, such as exercise;

e) 'stall' means a small enclosure with three sides, usually a feed-rack and lateral separations, where one or two animals may be kept tethered.

1	**THE PHYSICAL FACILITIES**
1.1	Functions and general design

1.1.1   Any facility should be so constructed as to provide a suitable environment for the species housed. It should also be designed to prevent access by unauthorised persons.

Facilities that are part of a larger building complex should also be protected by proper building measures and arrangements that limit the number of entrances and prevent unauthorised traffic.

1.1.2   It is recommended that there should be a maintenance programme for the facilities in order to prevent any defect of equipment.

1.2   **Holding rooms**

1.2.1   All necessary measures should be taken to ensure regular and efficient cleaning of the rooms and the maintenance of a satisfactory hygienic standard. Ceilings and walls should be damage-resistant with a smooth, impervious and easily washable surface. Special attention should be paid to junctions with doors, ducts, pipes and cables. Doors and windows, if any, should be constructed or protected so as to keep out unwanted animals. Where appropriate, an inspection window may be fitted in the door. Floors should be smooth, impervious and have a non-slippery, easily washable surface which can carry the weight of racks and other heavy equipment without being damaged. Drains, if any should be adequately covered and fitted with a barrier which will prevent animals from gaining access.

1.2.2   Rooms where the animals are allowed to run freely should have walls and floors with a particularly resistant surface material to stand up to the heavy wear and tear caused by the animals and the cleaning process. The material should not be detrimental to the health of the animals and be such that the animals cannot hurt themselves. Drains are desirable in such rooms. Additional protection must be given to any equipment or fixtures so that they may not be damaged by the animals or hurt the animals themselves. Where outdoor exercise areas are provided measures should be taken when appropriate to prevent access by the public and animals.

1.2.3   Rooms intended for the holding of farm animals (cattle, sheep, goats, pigs, horses, poultry, etc) should at least conform with the standards laid down in the European Convention for the Protection of Animals kept for Farming Purposes and by national veterinary and other authorities.

1.2.4   The majority of holding rooms are usually designed to house rodents. Frequently such rooms may also be used to house larger species. Care should be taken not to house together species which are incompatible.

1.2.5   Holding rooms should be provided with facilities for carrying out minor experiments and manipulations, where appropriate.

1.3	**Laboratories and general and special purpose experiment rooms**

1.3.1   At breeding or supplying establishments suitable facilities for making consignments of animals ready for dispatch should be made available.

1.3.2   All establishments should also have available as a minimum laboratory facilities for the carrying out of simple diagnostic tests, post-mortem examinations, and/or the collection of samples which are to be subjected to more extensive laboratory investigations elsewhere.

1.3.3   Provision should be made for the receipt of animals in such a way that incoming animals do not put at risk animals already present in the facility, for example by quarantining. General and special purpose experiment room should be available for situations where it is undesirable to carry out the experiments or observations in the holding room.

1.3.4   There should be appropriate accommodation for enabling animals which are ill or injured to be housed separately.

1.3.5   Where appropriate, there should be provision for one or more separate operating rooms suitably equipped for the performance of surgical experiments under aseptic conditions. There should be facilities for post-operative recover where this is warranted.

1.4	**Service rooms**

1.4.1   Store rooms for food should be cool, dry, vermin and insect proof and those for bedding, dry, vermin and insect proof. Others materials, which may be contaminated or present a hazard, should be stored separately.

1.4.2   Store rooms for clean cages, instruments and other equipment should be available.

1.4.3   The cleaning and washing room should be large enough to accommodate the installations necessary to decontaminate and clean used equipment. The cleaning process should be arranged so as to separate the flow of clean and dirty equipment to prevent the contamination of newly cleaned equipment. Walls and floors should be covered with a suitably resistant surface material and the ventilation system should have ample capacity to carry away the excess heat and humidity.

1.4.4   Provision should be made for the hygienic storage and disposal of carcasses and animal waste. If incineration on the site is not possible or desirable, suitable arrangements should be made for the safe disposal of such material having regard to local regulations and by-laws. Special precautions should be taken with highly toxic or radioactive waste.

1.4.5   The design and construction of circulation areas should correspond to the standards of the holding rooms. The corridors should be wide enough to allow easy circulation of movable equipment.

2	**THE ENVIRONMENT IN THE HOLDING ROOMS and ITS CONTROL**

2.1   **Ventilation**

2.1.1   Holding rooms should have an adequate ventilation system which should satisfy the requirements of the species housed. The purpose of the ventilation system is to provide fresh air and to keep down the level of odours, noxious gases, dust and infectious agents of any kind. It also provides for the removal of excess heat and humidity.

2.1.2   The air in the room should be renewed at frequent intervals. A ventilation rate of 15-20 air changes per hour is normally adequate. However, in some circumstances, where stocking density is low, 8-10 air changes per hour may suffice or mechanical ventilation may not even be needed at all. Other circumstances may necessitate a much higher rate of air change. Recirculation of untreated air should be avoided. However, it should be emphasized that even the most efficient system cannot compensate for poor cleaning routines or negligence.

2.1.3   The ventilation system should be so designed as to avoid harmful draughts

2.1.4   Smoking in rooms where there are animals should be forbidden.

2.2   **Temperature**

2.2.1   Table A7.1 gives the range within which it is recommended that the temperature should be maintained. It should also be emphasized that the figures apply only to adult, normal animals. Newborn and young animals will often require a much higher temperature level. The temperature of the premises should be regulated according to possible changes in the animals' thermal regulation which may be due to special physiological conditions or to the effects of the experiment.

2.2.2   Under the climatic conditions prevailing in Europe it may be necessary to provide a ventilation system having the capacity both to heat and to cool the air supplied.

2.2.3   In user establishments a precise temperature control in the holding rooms may be required, because the environmental temperature is a physical factor which has a profound effect on the metabolism of all animals.

2.3   **Humidity**

Extreme variations in relative humidity (RH) have an adverse effect on the health and well-being of animals. It is therefore recommended that the RH level in holding rooms should be appropriate to the species concerned and should ordinarily be maintained at 55% + 10% Values below 40% and above 70% RH for prolonged periods should be avoided.

2.4	**Lighting**

In windowless rooms, it is necessary to provide controlled lighting both to satisfy the biological requirements of the animals and to provide a satisfactory working environment. It is also necessary to have a control of the intensity and of the light-dark cycle. When keeping albino animals, one should take into account their sensitivity to light (see also 2.6).

2.5	**Noise**

Noise can be an important disturbing factor in the animal quarters. Holding rooms and experiment rooms should be insulated against loud noise sources in the audible and the higher frequencies in order to avoid disturbances in the behaviour and the physiology of the animals. Sudden noises may lead to considerable change in organ functions but, as they are often unavoidable, it is sometimes advisable to provide holding and experiment rooms with a continuous sound of moderate intensity such as soft music.

2.6	**Alarm systems**

A facility housing a large number of animals is vulnerable. It is therefore recommended that the facility is duly protected by the installation of devices to detect fires and the intrusion of unauthorised persons. Technical defects or a breakdown of the ventilation system is another hazard which should cause distress and even the death of animals, due to suffocation and overheating or, in less serious cases, have such negative effects on an experiment that it will be a failure and have to be repeated. Adequate monitoring devices should therefore be installed in connection with the heating and ventilation plant to enable the staff to supervise its operation in general. If warranted, a stand-by generator should be provided for the maintenance of life support systems of the animals and lighting in the event of a breakdown or the withdrawal of supply. Clear instructions on emergency procedures should be prominently displayed. Alarms for fish tanks are recommended in case of failure of the water supply. Care should be taken to ensure that the operation of an alarm system causes as little disturbance as possible to the animals.

3	**CARE**

3.1	**Health**

3.1.1	The person in charge of the establishment should ensure regular inspection of the animals and supervision of the accommodation and care by a veterinarian or other competent person.

3.1.2	According to the assessment of the potential hazard to the animals, appropriate attention should be paid to the health and hygiene of the staff.

3.2	**Capture**

Wild and feral animals should be captured only by humane methods and by experienced persons who have a thorough knowledge of the habits and habitats of the animals to be caught. If an anaesthetic or any other drug has to be used in the capturing operation, it should be administered by a veterinarian or other competent person. Any animal which is seriously injured should be presented as soon as possible to a veterinarian for treatment. If the animal, in the opinion of the veterinarian, can only go on living with suffering or pain it should be killed at once by a humane method. In the absence of a veterinarian, any animal which may be seriously injured should be killed at once by a humane method.

3.3   **Packing and transport conditions**

All transportation is undoubtedly, for the animals, a stressful experience, which should be mitigated as far as possible. Animals should be in good health for transportation and it is the duty of the sender to ensure that they are so. Animals which are sick or otherwise out of condition should never be subjected to any transport which is not necessary for therapeutic or diagnostic reasons. Special care should be exercised with female animals in an advanced state of pregnancy. Female animals which are likely to give birth during the transport or which have done so within the preceding forty-eight hours, and their offspring, should be excluded from transportation. Every precaution should be taken by sender and carrier in packing, stowing and transit to avoid unnecessary suffering through inadequate ventilation exposure to extreme temperatures, lack of feed and water, long delays, etc. The receiver should be properly informed about the transport details and documentary particulars to ensure quick handling and reception in the place of arrival. It is recalled that, as far as international transport of animals is concerned. Directives 77/489/EEC and 81/389/EEC apply; strict observance of national laws and regulations as well as of the regulations for live animals of the International Air Transport Association and the Animal Air Transport Association is also recommended.

3.4   **Reception and unpacking**

The consignments of animals should be received and unpacked without avoidable delay. After inspection, the animals should be transferred to clean cages or pens and be supplied with feed and water as appropriate. Animals which are sick or otherwise out of condition should be kept under close observation and separately from other animals. They should be examined by a veterinarian or other competent person as soon as possible and, where necessary, treated. Animals which do not have any chance to recover should be killed at once by a humane method. Finally, all animals received must be registered and marked in accordance with the provisions of Articles 17, 18, 19 (5) of the Directive Transport boxes should be destroyed immediately if proper decontamination is impossible.

3.5   **Quarantine, isolation and acclimatisation**

3.5.1   The objects of quarantine are:

                a) to protect other animals in the establishment;

                b) to protect man against zoonotic infection;

                c) to foster good scientific practice.

Unless the state of health of animals introduced into an establishment is satisfactory, it is recommended that they should undergo a period of quarantine. In some cases, that of rabies, for example, this period may be laid down in the national regulations of the Member State. In others, it will vary and should be determined by a competent person, according to the circumstances, normally the veterinarian appointed by the establishment (see also Table A7.2).

Animals may be used for experiments during the quarantine period as long as they have become acclimatised to their new environment and they present no significant risk to other animals or man.

3.5.2   It is recommended that facilities should be set aside in which to isolate animals showing signs of or suspected of ill-health and which might present a hazard to man or to other animals.

3.5.3   Even when the animals are seen to be in sound health it is good husbandry for them to undergo a period of acclimatisation before being used in an experiment. The time required depends on several factors, such as the stress to which the animals have been subjected which in turn depends on several factors such as the duration of the transportation and the age of the animal. This time shall be decided by a competent person.

3.6   **Caging**

3.6.1   It is possible to make a distinction between two broad systems of housing animals.

Firstly, there is the system found in breeding, supplying and user establishments in the bio-medical field designed to accommodate animals such as rodents, rabbits, carnivores, birds and non-human primates, sometimes also ruminants, swine and horses. Suggested guidelines for cages, pens, runs and stalls suitable for such facilities are presented in Tables A7.3 to A7.13. Supplementary guidance on minimum cage areas is found in Figures A7.1 to A7.7. Furthermore, a corresponding guidance for the appraisal of the stocking density in cages is presented in Figures A7.8 to A7.12.

Secondly, there is the system frequently found in establishments conducting experiments only on farm or similar large animals. The facilities in such establishments should not be less than those required by current veterinary standards.

3.6.2   Cages and pens should not be made out of material that is detrimental to the health of the animals, and their design should be such that the animals cannot injure themselves and, unless they are disposable, they should be made from a resistant material adapted to cleaning and decontamination techniques. In particular, attention should be given to the design of cage and pen floors which should vary according to the species and age of the animals and be designed to facilitate the removal of excreta.

3.6.3   Pens should be designed for the well-being of the species. They should permit the satisfaction of certain ethological needs (for example the need to climb, hide or shelter temporarily) and be designed for efficient cleaning and freedom from contact with other animals.

3.7	**Feeding**

3.7.1   In the selection, production and preparation of feed, precautions should be taken to avoid chemical, physical and microbiological contamination. The feed should be packed in tight, closed bags, stamped with the production date when appropriate. Packing, transport and storing should also be such as to avoid contamination, deterioration or destruction. Store rooms should be cool, dark, dry, and vermin and insect proof. Quickly perishable feed like greens, vegetables, fruit, meat, fish, etc should be stored in cold rooms, refrigerators or freezers.

     All feed hoppers, troughs or other utensils used for feeding should be regularly cleaned and if necessary sterilized. If moist feed is used or if the feed is easily contaminated with water, urine, etc, daily cleaning is necessary.

3.7.2   The feed distribution process may vary according to the species but it should be such as to satisfy the physiological needs of the animal. Provision should be made for each animal to have access to the feed.

3.8	**Water**

3.8.1   Uncontaminated drinking water should always be available to all animals. During transport, it is acceptable to provide water as part of a moist diet. Water is however a vehicle of micro-organisms and the supply should therefore be so arranged that the hazard involved in minimised. Two methods are in common use, bottles and automatic systems.

3.8.2   Bottles are often used with small animals like rodents and rabbits. When bottles are used, they should be made from translucent material in order to enable their contents to be monitored. The design should be wide-mouthed for easy and efficient cleaning, and, if plastic material is used, it should not be leachable. Caps, stoppers and pipes should also be sterilisable and easy to clean. All bottles and accessories should be taken to pieces, cleaned and sterilised at appropriate and regular periods. It is preferable that the bottles should be replaced by clean, sterilised ones rather than be refilled in the holding rooms.

3.8.3   Automatic drinking systems should be regularly checked, serviced and flushed to avoid accidents and the spread of infections If solid-bottom cages are used, care should be taken to minimise the risk of flooding. Regular bacteriological testing of the system is also necessary to monitor the quality of the water.

3.8.4   Water received from public waterworks contains some micro-organisms which are usually considered to be harmless unless one is dealing with microbiologically defined animals. In such cases, the water should be treated. Water supplied by public waterworks is usually chlorinated to reduce the growth of micro-organisms. Such chlorination is not always enough to keep down the growth of certain potential pathogens, as for example Pseudomonas. As an additional measure the level of chlorine in the water could be increased or the water could be acidified to achieve the desired effect.

3.8.5   In fishes, amphibians and reptiles, tolerance for acidity, chlorine and many other chemicals differs widely from species to species. Therefore, provision should be made to adapt the water supply for aquariums and tanks to the needs and tolerance limits of the individual species.

| 3.9 | **Bedding** |

Bedding should be dry, absorbent, non-dusty, non-toxic and free from infectious agents or vermin, or any other form of contamination. Special care should be taken to avoid using sawdust or bedding material derived from wood which has been treated chemically. Certain industrial by-products or waste, such as shredded paper, may be used.

| 3.10 | **Exercising and handling** |

| 3.10.1 | It is advisable to take every possible opportunity to let animals take exercise. |

| 3.10.2 | The performance of an animal during an experiment depends very much on its confidence in man, something which has to be developed. The wild or feral animal will probably never become an ideal experimental animal. It is different with the domesticated animal born and raised in contact with man. The confidence once established should however be preserved. It is therefore recommended that frequent contact should be maintained so that the animals become familiar with human presence and activity. Where appropriate, time should be set aside for taking, handling and grooming. The staff should be sympathetic, gentle and firm when associating with the animals. |

| 3.11 | **Cleaning** |

| 3.11.1 | The standard of facility depends very much on good hygiene. Clear instructions should be given for the changing of bedding in cages and pens. |

| 3.11.2 | Adequate routines for the cleaning, washing, decontamination and, when necessary, sterilisation of cages and accessories, bottles and other equipment should be established. A very high standard of cleanliness and order should also be maintained in holding, washing and storage rooms. |

| 3.11.3 | There should be regular cleaning and where appropriate, renewal of the material forming the ground surface in outdoor pens, cages and runs to avoid them becoming a source of infection and parasite infestation. |

| 3.12 | **Humane killing of animals** |

| 3.12.1 | All humane methods of killing animals require expertise which can only be attained by appropriate training. |

| 3.12.2 | A deeply unconscious animal can be exsanguinated but drugs which paralyse muscles before unconsciousness occurs, those with curariform effects and electrocution without passage of current through the brain, should not be used without prior anaesthesia. |

Carcass disposal should not be allowed until *rigor mortis* occurs.

Species or groups of species	Optimal range in °C
Non-human New World primates	20-28
Mouse	20-24
Rat	20-24
Syrian hamster	20-24
Gerbil	20-24
Guinea pig	20-24
Non-human Old World primates	20-24
Quail	20-24
Rabbit	15-21
Cat	15-21
Dog	15-21
Ferret	15-21
Pountry	15-21
Pigeon	15-21
Swine	10-24
Goat	10-24
Sheep	10-24
Cattle	10-24
Horse	10-24

Note: In special cases, for example when housing very young or hairless animals, higher room temepratures than those idnicated may be required.

Table A7.1 Guidelines for room temperature (animals kept in cages, pens or indoor runs).

Guidelines for local quarantine periods

Introductory note: For imported animals, all quarantine period should be subject ot the Member States' national regulations. In regard to local quarantine periods, the period should bedetermined by a competent person according to circumstances, normally a veterinarian appointed by the establishment.

Species	Days
Mouse	5 - 15
Rat	5 - 15
Gerbil	5 - 15
Guinea pig	5 - 15
Syrian hamster	5 - 15
Rabbit	20 - 30
Cat	20 - 30
Dog	20 - 30
Non-human primates	40 - 60

Table A7.2 Guidelines for local quarantine periods.

Species	Minimum cage floor area cm^2	Minimum cage height cm
Mouse	180	12
Rat	350	14
Syrian hamster	180	12
Guinea pig	600	18
Rabbit 1kg	1400	30
2kg	2000	30
3kg	2500	35
4kg	3000	40
5kg	3600	40

Note: 'Cage height' means the vertical distance between the cage floor and the uper horizontal part of the lid or cage.

When designing experiments, consideration should be given to the potential growth of the animals to ensure adequate room accoding to this table in all phases of the experiments.

See also Figures A7.1 to 5 and A7.8 to 12.

Table A7.3 Guidelines for caging small rodents and rabbits (in stock and during experiments).

Species	Minimum cage floor area for mother and litter (cm^2)	Minimum cage height (cm)
Mouse	200	12
Rat	800	14
Syrian hamster	650	12
Guinea pig	1200	18
Guinea pig in harems	1000 per adult	18

Table A7.4 Guidelines for caging small rodents in breeding.

Weight of doe (kg)	Minimum cage floor area per doe and litter (m^2)	Minimum cage height cm	Minimum nest box floor (m^2)
1	0.30	30	0.10
2	0.35	30	0.10
3	0.40	35	0.12
4	0.45	40	0.12
5	0.50	40	0.14

Note: for definition of 'cage height' see note to Table A7.3.

The minimum cage floor area per doe and litter includes the areaof the nest box floor.

See also Figure A7.6.

Table A7.5 Guidelines for caging breeding rabbits.

Weight of cat (kg)	Minimum cage floor area per cat (m^2)	Minimum cage height cm	Minimum cage floor area per queen and litter (m^2)	Minimum pen floor area per queen and litter (m^2)
0.5 - 1	0.2	50	-	-
1 - 3	00.3	50	0.58	2
3 - 4	0.4	50	0.58	2
4 - 5	0.6	50	0.58	2

Note: The housing of cats in cages should be strictly limited.. Cats confined in this way should be let out for exercising at least once a day, where it does not interfere with the experiment. Cat pens should be equipped with dirt trays, ample shelf room for resting and objects suitable for climbing and claw-trimming.

'Cage height' means the vertical distance between the highest point on the floor and the lowest point in the top of the cage.

Table A7.6 Guidelines for housing cats (during experiments and breeding).

Height of dog to point of shoulder cm	Minimum cage floor area per dog m^2	Minimum height of cage cm
30	0.75	60
40	1.00	80
70	1.75	140

Note: Dogs should not be kept in cages any longer than is absolutely necessary for the purpose of the experiment. Caged dogs should be let out for exercise at least once a day unless it is incompatible with the purpose of the experiment. A time- limit should be set beyond which a dog should not be confined without daily exercise. Exercise areas should be large enough to allow the dog freedom of movement. Grid floors should not be used in dog cages unless the experiment requires it.

In the light of the great differences in height and the limited interdependence of height and weight of various breeds of dogs, the cage height should be based on the body height to the shoulder of the individual animal. As a general rule the minimum cage height should be twice the height to the shoulder.

For definition of 'cage height', see note to Table A7.6.

Table A7.7 Guidelines for housing dogs in cages (during experiments).

Weight of dog kg	Minimum pen floor area per dog m^2	Minimum adjacent exercise area per dog	
		up to 3 dogs m^2	more than 3 dogs m^2
6	0.5	0.5 (1.0)	0.5 (1.0)
6 - 10	0.7	1.4 (2.1)	1.2 (1.9)
10 - 20	1.2	1.6 (2.8)	1.4 (2.6)
20 - 30	1.7	1.9 (3.6)	1.6 (3.3)
30	2.0	2.0 (4.0)	1.8 (3.8)

Note: Figures in brackets give the total area per dog, that is, the pen floor area plus the adjacent exercise area. Dogs kept, permanently outdoors should have access to a sheltered place to find protection against unfavourable weather conditions. Where dogs are housed on grid floors, a solid area should be provided for sleeping. Grid floors should not be used unless the experiment requires it. Partitions between pens should be such as to prevent dogs from injuring each other.

All pens should have adequate drainage.

Table A7.8 Guidelines for housing dogs in pens (in stock and during experiments and breeding).

Introductory note:Because of the wide variations in sizes and characteristics of primates, it is especially important to match the shape and internal fittings as well as the dimensions of their cages to their particular needs. The total volume of the cage is just as important to primates as the floor area. As a general principle, the height of a cage, at least for apes and other simians, should be its greatest dimension. Cages should be high enough at least to allow the animals to stand up erect. The minimum cage height for brachiators should be such as to allow them to swing in full extension from the ceiling without their feet touching the cage floor. Where appropriate, perches should be fitted to allow the primates to use the upper part of the cage.

Compatible primates may be kept two to a cage. Where they cannot be kept in pairs, their cages should be so placed that they can see one another, but it should also be possible to prevent this when required.

Subject to these observations, the following table constitutes a general guideline for caging the groups of species most commonly used (superfamilies Ceboidea and Cercopithecoidea).

Weight of primate	Minimum cage floor area for one or two animals	Minimum cage height
kg	m2	cm
1	0.25	60
1 - 3	0.35	75
3 - 5	0.50	80
5 - 7	0.70	85
7 - 9	0.90	90
9 - 15	1.10	125
15 - 25	1.50	125

Note: for definition of 'cage height' see note to TableA7. 6.

Table A7.9 Guidelines for caging non-human primates (in stock and during experiments and breeding).

Weight of pig	Minimum cage floor area per pig	Minimum cage height
kg	m^2	cm
5 - 15	0,35	50
15 - 25	0,55	60
25 40	0,80	80

Note: The table would also apply to piglets. Pigs should not be kept in cages unless absolutely necessary for the purpose of the experiment and then only for a minimum period of time.

For definition of 'cage height' see note to Table A7.6.

Table A7.10 Guidelines for caging pigs (in stock and during experiments).

Species and weights	Minimum pen floor area	Minimum pen length	Minimum pen partition height	Minimum pen floor area for groups	Minimum length of feed rack per head
(kg)	(m^2)	(m)	(m)	(m^2/animal)	(m)
Pigs 10 - 30	2	1.6	0.8	0.2	0.20
30 - 50	2	1.8	1.0	0.3	0.25
50 - 100	3	2.1	1.2	0.8	0.30
100 - 150	5	2.5	1.4	1.2	0.35
> 150	5	2.5	1.4	2.5	0.40
Sheep F	1.4	1.8	1.2	0.7	0.35
Goats F	1.6	1.8	2.0	0.8	0.35
Cattle 60	2.0	1.1	1.0	0.8	0.30
60 - 100	2.2	1.8	1.0	1.0	0.30
100 - 150	2.4	1.8	1.0	1.2	0.35
150 - 200	2.5	2.0	1.2	1.4	0.40
200 - 400	2.6	2.2	1.4	1.6	0.55
> 400	2.8	2.2	1.4	1.8	0.65
Adult horses	13.5	4.5	1.8	-	-

Table A7.11 Guidelines for accommodating farm animals in pens (in stock and during experiments in user establishments).

Species and weights	Minimum stall area	Minimum stall length	Minimum stall partition height
(kg)	(m^2)	(m)	(m)
Pigs 100 - 150	1.2	2.0	0.9
> 50	2.5	2.5	1.4
Sheep F	0.7	1.0	0.9
Goats F	0.8	1.0	0.9
Cattle 60 - 100	0.6	1.0	0.9
100 - 150	0.9	1.4	0.9
150 -200	1.2	1.6	1.4
200 - 350	1.8	1.8	1.4
350 - 500	2.1	1.9	1.4
> 500	2.6	2.2	1.4
Adult horses	4.0	2.5	1.6

Note: Stalls should be sufficiently wide to allow an animal to lie comfortably.

Table A7.12 Guidelines for accommodating farm animals in stalls (in stock and during experiments in user establishments).

Species and weights	Minimum area for one bird	Minimum area for 2 birds	Minimum area for 3 birds or more	Minimum cage height	Minimum length of feed trough per bird
(g)	(cm^2)	(cm^2/bird)	(cm^2/bird)	(cm)	(cm)
Chickens 100 - 300	250	200	150	25	3
300 - 600	500	400	300	35	7
600 - 1200	1000	600	450	45	10
1200 - 1800	1200	700	550	45	12
1800 - 2400	1400	850	650	45	12
(Adult males)					
> 2400	1800	1200	1000	60	15
Quails 120 - 140	350	250	200	15	4

Note: 'Area' means the product of cage length and cage width measured internally and horizontally. Not the product of the floor length and floor width.

For definition of 'cage height' see note to Table A7.6.

Mesh size in grid floors should not be greater than 10 x 10 mm for young chicks, and 25 x 25 mm for pullets and adults. The wire thickness should be at the least 2 mm. The sloping gradient should not exceed 14% (8°). Water troughs should be of the same length as the feed troughs. If nipples or cups are provided, each bird should have access to two. Cages should be fitted with perches and allow birds in single cages to see each other.

Table A7.13 Guidelines for caging birds (in stock and during experiments in user establishments).

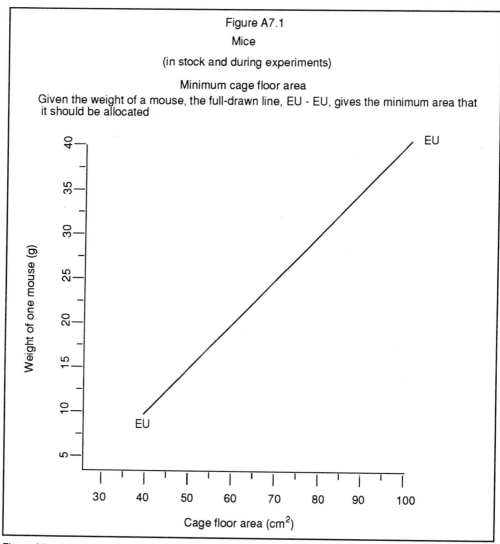

Figure A7.1 Mice (in stock and during experiments). Minimum cage floor area.

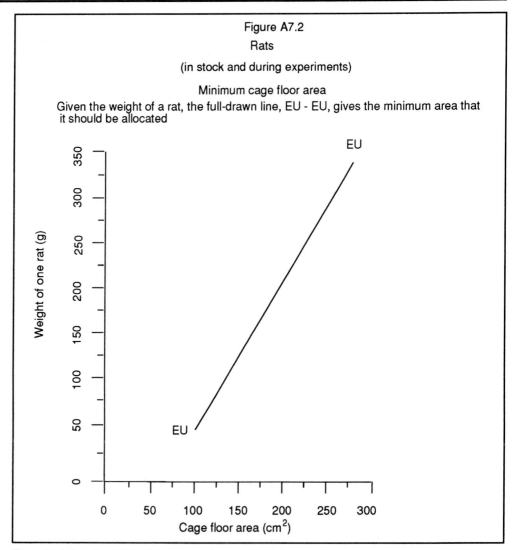

Figure A7.2 Rats (in stock and during experiments). Minimum cage floor area.

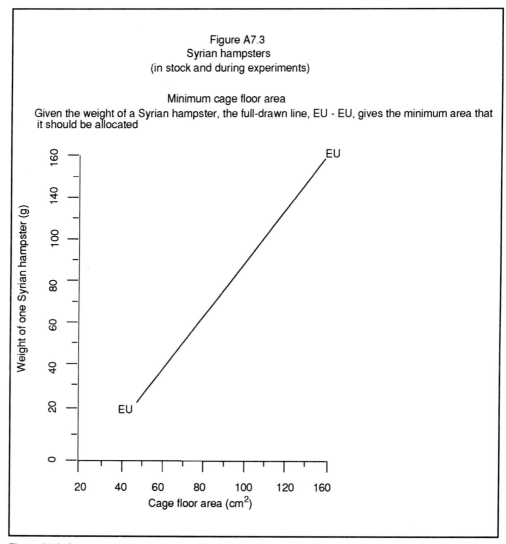

Figure A7.3 Syrian hamsters (in stock and during experiments). Minimum cage floor area.

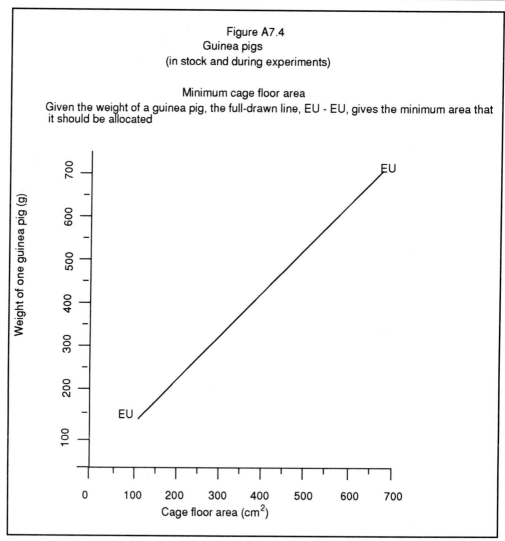

Figure A7.4 Guinea pigs (in stock and druing experiments). Minimum cage floor area.

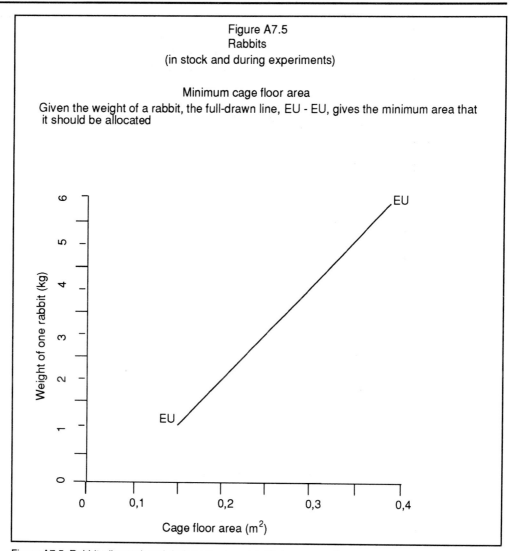

Figure A7.5
Rabbits
(in stock and during experiments)

Minimum cage floor area
Given the weight of a rabbit, the full-drawn line, EU - EU, gives the minimum area that it should be allocated

Figure A7.5 Rabbits (in stock and during experiments). Minimum cage floor area.

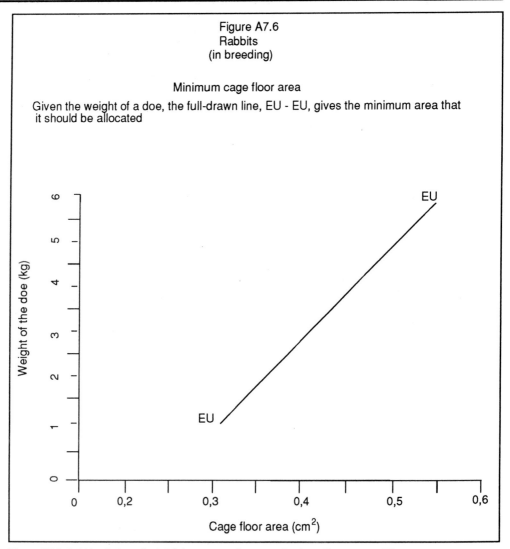

Figure A7.6
Rabbits
(in breeding)

Minimum cage floor area

Given the weight of a doe, the full-drawn line, EU - EU, gives the minimum area that it should be allocated

Figure A7.6 Rabbits (in breeding). Minimum cage floor area for doe with unweaned litter.

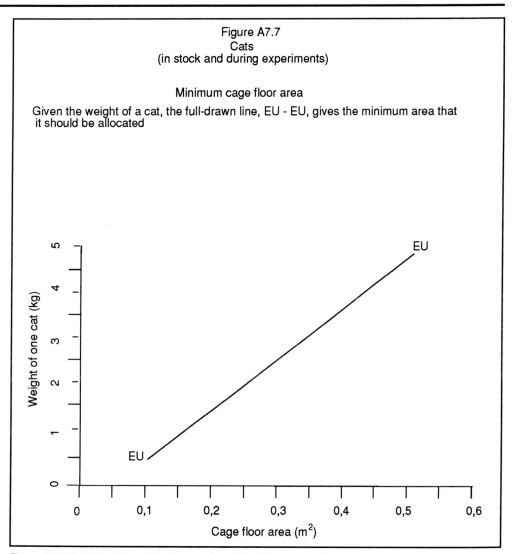

Figure A7.7 Cats (in stock and during experiments). Minimum cage floor area.

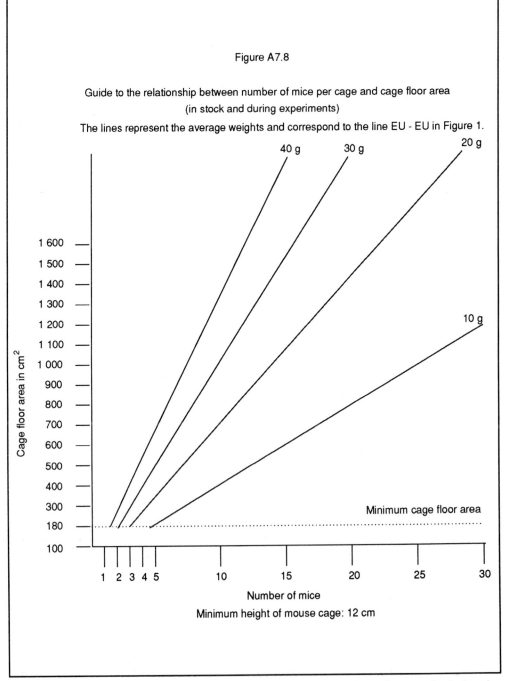

Figure A7.8 Guide to the relationship between number of mice per cage and cage floor area (in stock and during experiments).

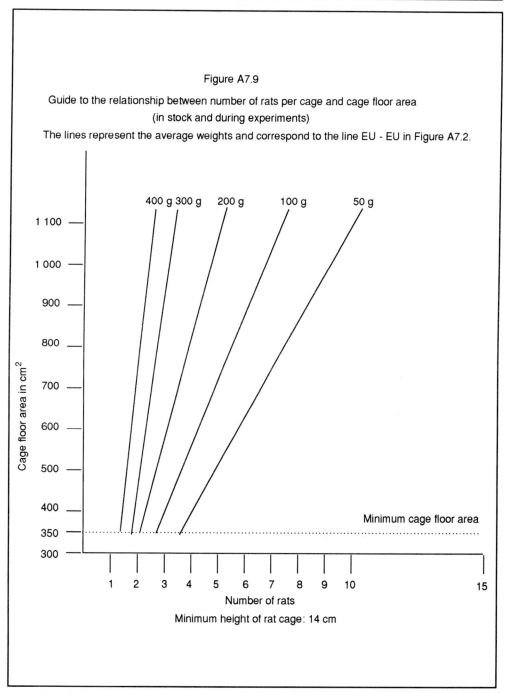

Figure A7.9

Guide to the relationship between number of rats per cage and cage floor area
(in stock and during experiments)
The lines represent the average weights and correspond to the line EU - EU in Figure A7.2.

Figure A7.9 Guide to the relationship between number of rats per cage and cage floor area (in stock and during experiments).

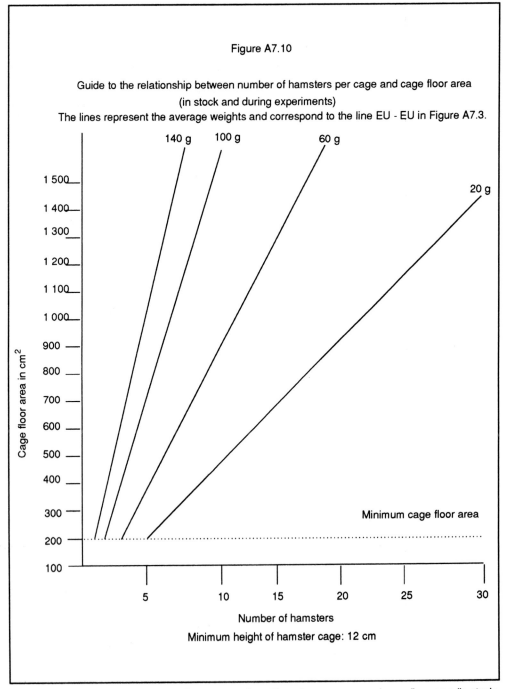

Figure A7.10

Guide to the relationship between number of hamsters per cage and cage floor area
(in stock and during experiments)
The lines represent the average weights and correspond to the line EU - EU in Figure A7.3.

Figure A7.10 Guide to the relationship between number of hamsters per cage and cage floor area (in stock and during experiments).

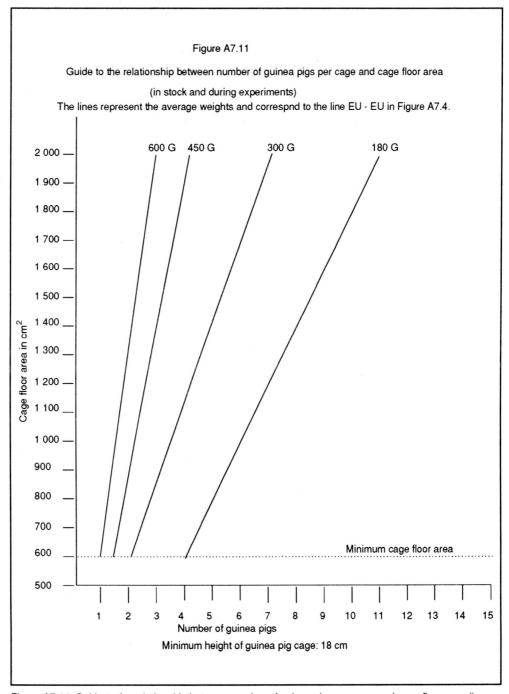

Figure A7.11 Guide to the relationship between number of guinea pigs per cage and cage floor area (in stock and during experiments).

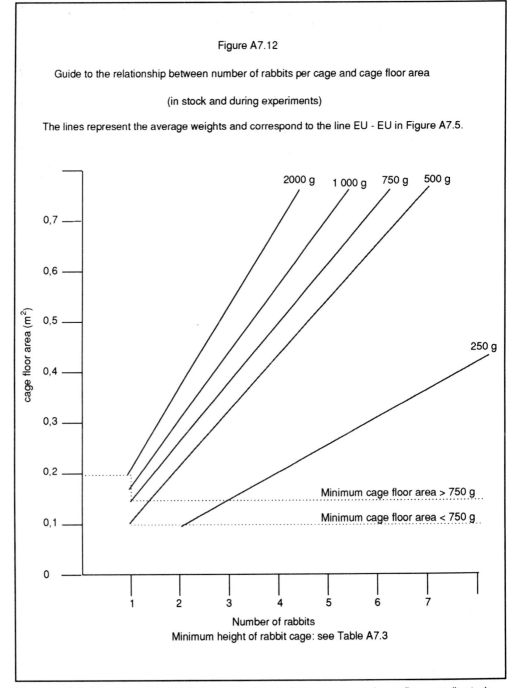

Figure A7.12

Guide to the relationship between number of rabbits per cage and cage floor area

(in stock and during experiments)

The lines represent the average weights and correspond to the line EU - EU in Figure A7.5.

Figure A7.12 Guide to the relationship between number of rabbits per cage and cage floor area (in stock and during experiments).

Radiation health and safety

Appendix 8.1

Radiation health and safety

8.1 Introduction

For the purposes of this chapter it is assumed that the reader has some basic background knowledge about atoms, electrons, protons, neutrons, molecules, elements etc. However, we have provided a summary of the basic facts concerning radiation in an appendix at the end of this chapter. We also briefly outline the effects of radiation on cells and explain the use of units used in radiobiology in the appendix. The main text confines itself to the issues and regulations concerning the protection of workers. We will also consider sources of radiation and the laboratory uses of radioactive nuclides and the disposal of radioactive materials. A glossary of terms is given in the appendix at the end of the chapter.

8.2 Radiation protection

8.2.1 ICRP and the EC

In terms of establishing the levels of protection of individuals and populations against radiation, a number of authorities have been established. By far the most important of these is the International Commission on Radiological Protection (ICRP) formed in 1928. It was established to make recommendations on all aspects of radiation protection. These recommendations were considered by national radiological protection authorities who made recommendations to national law making bodies. The actual titles of these national bodies authorities vary but we can represent the general flow of information in the following way.

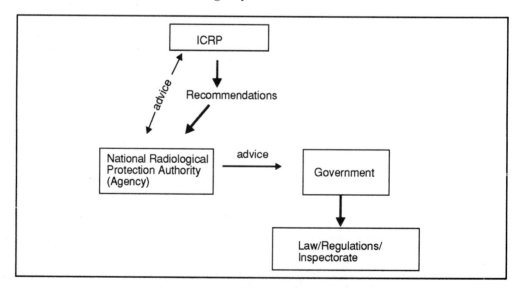

Thus, long before the EC was established as a superanational body, the activities of the ICRP led to a general harmonisation of standards and procedures between countries.

The formation of the EC has however led to a general increase in the harmonisation of standards and procedures.

European legislation on protection against ionising radiation

The earliest trans-European State legislation which concerned radiation was in 1959 and was a European Atomic Energy Community (EAEC), Council Directive entitled "Directives laying down the basic standards for the protection of the health of workers and the general public against the dangers arising from ionizing radiations". Apart from minor amendments in 1976 (Directive 6/579/Euratom) no major review occurred until 1980 despite the fact that in the intervening years since 1959 there has been considerable development of scientific knowledge concerning the hazards of radiation and radiation protection. The 1980 review was in the form of a Council Directive (80\836\Euratom) entitled "Amending the directives laying down the basic safety standards for the health protection of the general public and workers against the dangers of ionising radiation". A further amendment was made in 1984 (Council Directive 84/467/Euratom) entitled "Basic safety standards for the health protection of the general public and workers against the dangers of ionising radiation". It was in response to these two directives (80/836/Euratom and 84/46/Euratom) that, for example, the Ionising Radiations Regulations 1985 were made in the UK.

Further EC legislation relating to radiation has been made with regard to certain areas such as:

- protection of the public against radon;

- controls regarding shipments of radioactive materials;

- supervision of radioactive waste;

- arrangements for the early exchange of information in the event of a radiological emergency.

One piece of legislation of interest to medical biotechnologists is Council Directive (89/343/EEC) entitled "extending the scope of directives 65/65 EEC and 75/319/EEC and laying down additional provisions for radiopharmaceuticals". This legislation concerns the harmonisation of the laws in EC Member States and lays down the procedures and requirements of authorisations with respect of the use and marketing of radiopharmaceutical compounds. This legislation relates to previous laws (Council Directive 87/33EEC) on the placing on the market of high technology medicinal products, particularly those derived from biotechnology.

Since the ICRP is the principle standard setting body for radiation protection and many countries throughout the world have adopted their recommendations, we will predominantly focus attention on their recommendations.

8.2.2 The general aims of radiation protection

The aim of radiation protection should be to prevent detrimental non-stochastic effects (ie keep doses below threshold levels) and to limit the probability of stochastic effects to levels deemed to be acceptable. Three key recommendations are:

- no practice shall be adopted unless its introduction produces a positive net benefit;

- all exposures shall be kept as low as reasonably achievable, economic and social factors being taken into account;

- the exposure of individuals shall not exceed the limits recommended for the appropriate circumstances by the Commission.

8.2.3 Recommended dose equivalent limits

ICRP26 (1977) recommended that the maximum permissible effective dose equivalent should be: for radiation workers an annual dose equivalent limit for uniform irradiation of the whole body of 50 mSv (although individual parts of the body may receive higher doses). For members of the general public the level is 5 mSv per year.

> In 1987, following a reassessment of the Hiroshima and Nagasaki data (effects following nuclear explosions) a number of national radiation protection agencies (for example the UK National Radiation Protection Board) recommended stricter controls because the data suggested higher risk factors than hitherto realised. In the UK this was set at 15 mSv effective dose equivalent per year for radiation workers and for members of the general public, 0.5 mSv per year.

In 1990, the ICRP recommended that for radiation workers, the dose should be 100 mSv in 5 years with an overriding limit of 50mSv in any single year. For members of the public this should be 1mSv per year or 'in special cases' for it to be higher than this in any one year providing it is no more than 1mSv per year averaged over 5 years.

Despite these upper limits, there is an over-riding philosophy that doses should be kept as low as is readily achievable.

8.2.4 Secondary limits (annual limits of intake - ALI)

It is relatively easy to measure sources of radiation external to the body and their potential hazard and potential dose equivalent assessed. However, if radionuclides enter the body either by being breathed in, ingested or absorbed through the skin the subsequent assessment of dose is greatly complicated due to the behaviour of the different types of radionuclide since:

> i) the distribution of different radionuclides in the body is different;

> ii) the distribution is often uneven;

> iii) different radionuclides are metabolised differently.

To meet these difficulties, the ICRP publishes a list of annual limits of intake (ALI) by inhalation or ingestion. These secondary limits are designed to ensure that the internal dose does not exceed the recommended dose equivalent limit. The annual limit of intake applies to workers occupationally exposed to radiation, and it is derived on the basis of certain assumptions. For example, the values quoted are for a "hypothetical" reference human of average size, weight and metabolism living in an average climate. Clearly the ALI must be adjusted to take account of differences in biological characteristics. Moreover, some radiation doses cannot be easily measured directly and therefore have to be estimated from models.

8.3 Sources of radiation

We can broadly divide the sources of radiation into two types:

* natural sources;

* artificial sources.

We will briefly consider each in turn.

8.3.1 Natural sources of radiation

Radiation of natural origin is always present in the environment arising from four main sources which are:

Cosmic rays

About 10% of human exposure comes from this source. Despite the atmosphere acting as a partial shield, cosmic rays from the sun and outer space still reach the earths' surface. Passengers travelling in aeroplanes receive a greater radiation dose than those travelling at sea level.

Rocks, soil and natural building materials

Soils and rock contain radioactive materials such as uranium-238, thorium-232 and potassium-40. The magnitude of the radiation dose from these sources varies considerably depending on the geology of the land and other factors. Most building materials are extracted from the earth, so these can also be slightly radioactive. About 14% of natural exposure comes from this source.

Internal exposure from ingested radionuclides

Potassium - 40, lead-210, polonium-210 and other naturally occurring nuclides (from the ground or produced by cosmic rays) are present in the average diet and are therefore ingested. Approximately 12% of exposure arises from "internal" exposure.

Exposure to radon and its decay products

Naturally occurring uranium, present in certain rocks (eg. granite) produces radon gas as it decays. As a gas, radon diffuses readily through rock and soil and can be trapped in houses, especially those which are not well ventilated. Radon-222 decays (half life 3.82 d) into polonium-218 which decays within minutes into lead-214, bismuth-214 and polonium-214. These decay products unlike radon are solids and are known collectively as "radon daughters". They decay (more slowly) into lead-210, bismuth-210 and polonium-210. Radon and radon daughters attached to dust particles can be breathed into the lungs. Since many of these nuclides are alpha-emitters, the radiation dose can be extremely localised and difficult to calculate as a dose equivalent to the whole body. A similar gas (known as thoron or radon-220) is produced from thorium decay. Together, radon and thoron may contribute up to 51% of the average total exposure in the population. However, exposure varies considerably depending on the local geology and how well individual houses or workplaces are ventilated.

In Europe, the four natural sources of radiation give a combined average dose equivalent of about 2.2 mSv per year.

8.3.2 Artificial sources of radiation

There are three main artificial sources of radiation to which we are all exposed. These include:

Medical procedures

Dental, chest and limb X-rays, and diagnostic procedures using radioisotopes all give rise to radiation doses to the body. Radiotherapy (eg in treating cancer) is more common in older than in younger people and may use quite high doses of radiation. However, for the population as a whole this represents only a small figure since radiotherapy is much less common than radiodiagnosis. An average figure of 12% of total annual exposure may be attributed to medical procedures.

Fallout from nuclear explosions and the Chernobyl nuclear accident

In Europe, exposure due to fallout from weapons testing is now less than one-tenth of the highest levels recorded which occurred through the 1960's. The main radionuclides from weapons testing are caesium-137, strontium-90 and carbon-14.

National Radiological Protection agencies assess radioactive fallout by sampling rain, food, milk and airborne dust. Levels have diminished considerably following the Test Ban Treaty of 1963.

Deposition of radioactivity released by the Chernobyl accident in 1986 was heaviest in the areas where it was raining at the time the radioactive cloud was passing overhead. Although some "hotspots" still remain, the exposure in Europe due to Chernobyl has declined significantly in recent years.

Total fallout from both weapons testing and Chernobyl contribute less than 1% to the total annual exposure of the average European citizen.

Radioactive discharges

Discharges from the nuclear industry occur routinely from nuclear power stations and plants which manufacture or reprocess fuel. Hospitals and Universities also use radioactive material and therefore produce radioactive waste of one kind or another. Three methods are used to deal with radioactive waste:

a)   It is discharged to the environment either as gas into the air, or as liquid into the sewers or the sea. Some establishments are allowed to discharge up to a certain limit of activity each year by these means. The limits are set nationally. In the UK for example this is done by registration under the Radioactive Substances Act (1960).

b)   It can be stored on site.

c)   It can be disposed of or stored in a place where it is unlikely to contaminate the environment such as a deep underground store.

On average, only 0.1% of a persons total radiation exposure is likely to be due to radioactive discharges. However, some individuals who live near some nuclear establishments could potentially receive higher exposures.

8.4 Laboratory uses of radiation

Radioisotopes have a wide range of uses in biotechnological activities. These include:

- isotope tracing;
- isotope dilution analysis;
- radioimmunassays;

- radioactivation analysis;
- radiochromatography;
- autoradiography.

We will briefly describe each of these.

8.4.1 Isotope tracing

Radio-labelled molecules can be used to follow or trace the metabolic pathways of substrates or specific portions of their structures. Examples include; the fate of CO_2 when it is fixed by the dark reaction in photosynthesis, the uptake of radio-labelled amino acids and their incorporation into protein, the fate of radio-labelled glucose during glycolysis and Krebs cycle etc. As well as studying precursor-product relationships, radio-labelled molecules can also be used to follow the rates of processes, particularly in intact organisms.

8.4.2 Isotope dilution analysis

Many biological molecules are difficult to isolate or purify in a quantitative fashion. Thus it is difficult to estimate their concentration. Isotopic dilution enables the researcher to pursue high purity at the expense of quantitative recovery in the secure knowledge that a quantitative measure of the compound in question can still be made. This is achieved in the following way. Assume that we wish to purify compound X from a system and to determine the amount of X in the system. If we add a known amount of radioactivity labelled X (designated X*), it is assumed that X* behaves chemically in the same way as X. We now carry out our purification step(s) to recover only a fraction of X* by our purification procedure, then we can assume that we have recovered a similar fraction of the unlabelled X originally in the system. If we also know the specific radioactivity (amount of radioactive/unit mass) of X* added and we determine the specific radioactivity of the compound recovered by the purification procedure, we can calculation the amount of unlabelled X in the system.

This is because the specific radioactivity of X becomes

$$\frac{\text{total radioactivity added}}{\text{amount of } X^* \text{ added} + \text{amount of } X \text{ in the system}}$$

Since we know or determine all values except the amount of X in the system, we can calculate this value. There are many variations to this method (eg isotope derivative analysis, double isotope derivative analysis, reverse isotope dilution analysis) which can be used for particular purposes. The general term for all these methods is 'saturation analysis'.

8.4.3 Radiommunoassay

In general, this assay is based upon competition between a known and standardised amount of radio-labelled "antigen" (eg thyroid hormone, antibiotic, steroid hormone, vitamin B_{12} etc whatever is being measured) and an unknown amount of the same (but non-radioactive) antigen in a given sample (eg body fluid). The relative amounts of radio-labelled and non-labelled antigen that bind with antibody in the reagent indicate the levels of antigen in the sample. For example, high levels of radioactivity in the antibody-bound fraction indicate a low level of the antigen in the sample whilst low levels of radioactivity in the antibody-bound fraction indicate a high level of the antigen in the sample. The approach used is a type of saturation analysis as mentioned above, but uses immunological methods to isolate or capture the analyte of interest. The principle radioisotopes used to label antigen are ^{125}I, ^{14}C or ^{3}H.

8.4.4 Radioactivation analysis

Elements within a sample can be made radioactive by irradiation, usually by neutron bombardment in a reactor. After irradiation the sample (along with a suitably prepared standard containing known amounts of the elements in question) is analysed for both nature and type (eg energy spectra) and amount of radioactivity emitted. The ratios of unknown (sample) to known (standard) for any element of interest can be measured and the amount of the element in question within the sample can be calculated. This method is highly sensitive and can enable the levels of trace elements (eg lead) to be measured in a single hair or in other body samples. Thus, the technique is widely used in forensic science as well as in environmental analysis (eg To measure the levels of pesticides in grain or arsenic in wine etc).

8.4.5 Radiochromatography

The detection and measurement of radioactive fractions in thin layer chromatography plates or in the eluates from various types of chromatography columns (GC, HPLC) can be very useful in helping to identify various metabolites, particularly if they are only present in minute amounts. We can also add radioactive compounds to samples to be chromatographed to act as reference 'labels' in otherwise unlabelled samples (see also Section 8.4.2).

8.4.6 Autoradiography

genetic-finger-printing

Labelled whole or parts of plants and animals, labelled tissues or small sections of tissue can all be analysed by holding the specimen in contact with a "film" or emulsion of light sensitive chemicals (eg an X-ray film) in darkness for a suitable length of time. The film is then developed and the anatomical distribution of radioisotope in the specimen determined by matching the pattern on the developed film with the original specimen. This method is also useful for detecting the position of specific types of DNA following separation using electrophoresis. A radio-labelled "probe" (which is short-length, single-stranded DNA with a specific "anti-sense" sequence) is used in order to bind (and therefore specifically detect) the DNA with the "sense" sequence of interest. Autoradiography of the resulting electrophoretogram reveals the positions of the DNA in question. This method has important applications in forensic science eg in genetic-fingerprinting.

Two major issues arise form those types of studies:

- the protection of workers and the public;
- the control of discharges.

8.5 Protection of workers and the public

A large number of Regulations are applied at national-level but within the EC, these are progressively being tailored to implement (at least in part) the provisions of the EC-Directive 80/836/Euratom as amended by Directive 84/467/Euratom. These lay down the basic safety standards for the health protection of the general public and workers against the dangers of ionising radiation. These standards are then interpreted into National Regulations. In Table 8.1, we have provided a summary of the UK Regulations as an example.

Part I (Interpretation and general - Regulations 1-5) defines the terms used and sets out the scope of the Regulations. Apart from certain exceptions it makes it a requirement that all employers (including self-employed persons) must notify the Helath and safety Executive of work with ionising radiation.

Part II (Dose limitation - Regulations 6 and 7) requires every employer to take all necessary steps to provide systems of work as will (so far as reasonably practicable) restrict the extent to which employees and other persons are exposed to ionising radiation, by means of engineering controls and design features inclcuding shielding, ventilation, containment, monitoring and warning devices. The employer must also ensure that no radioactive substance in the form of a sealed source is held or manipulated by hand unless the instantaneous dose rate to the skin is less than 75 mSv per hour. Unsealed radioactive sources must not be touched by hand. Employees must not eat, drink, smoke, take snuff or apply cosmetics in areas which have been designated as controlled. Part 2 also requires employers to impose limits (specified in Schedule 1 of the Regulations) on the doses of ionising radiation which their employees (or other persons) may receive in any calendar year.

Part III (Regulation of work with ionising radiation - Regulations 8-12) provide that areas in which persons are likely to receive more than specified doses of radiation be designated as controlled areas or supervised areas and restrict entry to all except specified persons under specified circumstances. Employers who are likely to receive more than specified doses of ionising radiation are required to be designated as classified persons.

Part III of the regulations also cover: the appointment of radiation protection advisers (to advise the employer on observance of the regualtions and on radiation health and safety matters); the appointment of radiation supervisors (to supervise work in controlled areas); the requirement to make written local rules for the conduct of work with ionising radiation and to ensure that such work is closely supervised and that adequate information, instruction and training is given to employees and other persons. (An example of local rules in radiation protection are shown in appendix I at the end of this chapter).

Part IV (Dosimetry and medical surveillance - Regulations 13-17) requires that all doses of ionising radiation received by classfied and certain other specified persons are assessed by one or more dosimetry services approved by the HSE. Records of doses must be made and kept (for at least 50 years) for each such person. Provision must also be made for immediate dose assessment to be made following an accident, occurrence or incident which is likely to result in a person being exposed to more than three-tenths of any relevant dose limit. Part IV of the Regulations also requires that certain employees be subject to medical surveillance. Employers are required to make approved arrangements for the protection of the health of individual employees. Suitable health records must be maintained.

Part V (Arrangements for the control of radioactive substances - Regulations 18-23) makes reference to the design and construction of any article containing a radioactive substance. They must be suitably designed, constructed, maintained and tested. Where a radioactive substance is used as a source of ionising radiation, it should wherever reasonably practicable, be in the form of a sealed source. Each employer should account for, and keep records of, the quantities and locations of radioactive substances. With regard to keeping radioactive substances, the employer must ensure they are kept in suitable receptacles in a suitable store. With regard to transport of radioactive substances, employers must ensure that it is kept in a suitable receptable and that suitable written information accompanies the radioactive substances so that the receiver knows the nature and quantity of the material in question. Part V is also concerned with the provision of properly maintained washing and changing faciltities in certain cases and that any respiratory protective equipment used in work with ionising radiation be approved for the purposes. All protective equipment must be regularly examined and properly maintained.

Part VI (Monitoring of ionising radiation - Regulation 24) concerns the monitoring of levels of radiation and contamination in controlled and supervised areas and provides for monitoring equipment to be properly examined and maintained.

Table 8.3 Summary of the UK Ionising Radiations Regulations 1985: Summary of main parts and sections.
(continued)

Part VII (Assessments and notifications - Regulations 25-31) requires that every employer who undertakes work with ionising radiation to make an assessment of the hazards that are likely to arise from the work. In the cases where more than specified quantities of radioactive substances are involved, an assessment report must be sent to the HSE. The employer must also, in certain circumstances, prepare contingency plans designed to protect workers and the public in the event of foreseeable accident, occurrence or incident. Part VII also requires the employer to investigate cases where radiation exposure has exceeded given levels and notify the HSE. A modified dose limit may be set for emplyees who have received an overexposure. The Regulations also require that incidents where more than specified quantities of radioactive substances escape or are lost or stolen, must be notified to the HSE.

Part VIII (Safety of articles and equipment - Regulations 32-34) deals with the duties of manufacturers and installers of articles for use in work with ionising radiation. They have a duty to ensure that such articles are designed, constructed and installed so as to restrict (as far as is reasonably practicable) exposure to ionising radiation. The employer must be provided with adequate information about the proper use, testing and maintenance of articles. The employer must ensure that any equipment or apparatus under his control is properly constructed, installed and maintained; malfunctions must be investigated immediately and the HSE notified. Similar duties are imposed on employers in relation to equipment used for medical exposures and includes defects where a person undergoing medical exposure receives a much greater dose of ionising radiation than was intended. The HSE must be notified of such confirmed incidents. Part VIII also prohibits the interference with sources of ionising radiation.

Part IX (Miscellaneous and general - Regulations 35-41) refers to aspects of exemption and defence on contravention of certain regulations. The Regulations also contain transitional and other incidental provisions which apply to offshore situations, and introduce MOD modifications. Also covered in this Part is the payment of fees to the HSE for medical surveillance (under Regulations 16 [3]).

Schedule 1 of the regulations refers to dose limits (part I, whole body; part II, individual organs and tissues; part III, lens of the eye; part IV, abdomen of a woman of reproductive capacity; part V, abdomen of a pregnant woman.

Schedule 2 relates to quantities of radionuclides (for purposes of notification, controlled areas, assessment report and notification of occurrences following incident).

Schedule 3 sets out 9 descriptions of work which do not require to be noitified under regulation 5(2) (ie. exemptions).

Schedule 4 lists the particulars that must be supplied in a notification under regulation 5(2) whilst **Schedule 5** lists additional particulars that the HSE may require.

Schedule 6 concerns the requirements of the designation of controlled areas with respect to external, internal and combined radiation levels and in relation to the short-lived daughters of radon-222.

Schedule 7 lists the particulars to be included in an assessment report whilst Schedule 8 lists additional particulars that the HSE may require.

Schedule 9 is a list of specifications which if satisfied allow work to be undertaken despite restrictions set out in Regulation 26. The title of Schedule 9 is "sealed sources to which Regulation 26 does not apply".

Schedule 10 lists previous legislation that is either revoked (part 1) or modified (part 2) as a consequence of enactment of "The ionising radiations regulations 1985".

Table 8.3b Summary of the UK Ionising Radiations Regulations 1985: Summary of main parts and sections.

Thus Table 8.1 shows some of the common features of national regulations. These may be summarised as:

- the duties of employers to protect employees (and others);

- the duties of employees;

- the need to instal appropriate features into facilities;

- to monitor exposure and keep records;

- to arrange for health surveillance;

- to report accidents;

- to record discharges (see Section 8.9);

- training, supervision and inspection.

The amount of radioactive substances held and used by an installation is also controlled. The prospective user applies for authorisation to an inspectorate regarding the particular quantities and types of isotopes to be used. This, of course, reflects the nature and extent of the study to be undertaken.

Generally authorisation to use small quantities of radioactive materials is more easy to gain.

controlled and supervised areas

Usually, in European states, a system exists for distinguishing between two types of laboratories. In the UK, for example, there are called controlled and supervised laboratories.

Laboratory areas where high levels of radioactivity are being employed and doses are likely to exceed three-tenths of any relevant dose limit for workers should always be designated as Controlled areas. Access to such areas must be restricted and the area must be demarcated. It is unusual for the radioactivity facilities within medical and biological science laboratories to be so designated. It is more likely that they would be classed as Supervised areas. A Supervised area is one in which radiation doses are not likely to exceed one-tenth but not exceeding three-tenths of any relevant dose limit for radiation workers.

Classified worker

It is possible that a part of a laboratory (eg a demarcated bench area, a side laboratory, a lockable storage cupboard or freezer) is designated as a localised Controlled area within a Supervised area. This may be necessary in order to store or dilute stock materials which have higher levels of radioactivity than that used in routine work. A person who works in such a Classified area must either be i) a Classified worker (ie a radionuclide worker who is likely to receive a radiation dose which exceeds three-tenths of any relevant dose limit) and must be over 18 years of age and be subject to annual medical examination by an Appointed Doctor, or ii) must work under a written Systems of Work that ensures that they do not exceed three-tenths of any relevant dose limit (which must be checked by dosimeter monitoring).

The following table (from UK Ionising Radiations Regulations, 1985) (Table 8.2) shows the limiting levels of activity of some radionuclides (ones commonly used in biological sciences) for exemption of the laboratory area as a Controlled area. Limits are shown in relation to both external and internal radiation levels.

Radionuclide	External radiation	Internal radiation	
		Air concentration	Surface concentration
		(m^{-3})	$(cm^{-2}$
^3H	50 MBq	200kBq	4 MBq
^{14}C	50 MBq	10kBq	10kBq
^{22}Na	5 MBq	3 kBq	2 kBq
^{40}K	50 MBq	2 kBq	1 kBq
^{32}P	5 MBq	2 kBq	2 kBq
^{35}S	50 MBq	9 kBq	20 kBq
^{45}Ca	50 MBq	3kBq	7 kBq
^{59}Fe	5 MBq	2 kBq	4 kBq

Table 8.2 Limits of various nuclids for exemption of a laboratory as a controlled area.

Implied in this system of Controlled and Supervised areas, is the requirement to maintain records and to registry facilities with the appropriate authorities. The radiation laboratories, therefore, need to have a clear management system. It is usual for organisations to appoint a "Safety Officer" with particular responsibilities for ensuring that all aspects of the Regulations are adhered to. Others take responsibility over a particular area or activity within the laboratories. Such individuals carry various titles (for example Radiation Protection Adviser, Radiation Protection Supervisor). Thus a chain of responsibilities is established along the following lines:

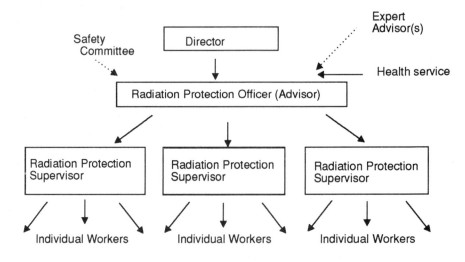

Although the "Director" is ultimately responsible for all aspects of radioactive work within an institution, all individuals in this chain are responsible for their own actions. Note that institutions may also call upon additional advisors and will be required to call upon health services to carry out health surveillance of individuals.

Codes of Practice

The Regulations, of the type described in Table 8.1, are often associated with a "Code of Practice". Such codes of practice are drawn up following widespread consultation with interested parties. The provisions of such codes of practice represent the most appropriate methods of complying with the regulatory requirements, including EC Directives 80/836 and 84/467. For example in the UK the Code of Practice entitled "The protection of persons against ionising radiation arising from any work activity" has been issued by the Health and Safety Executive. (Available from HMSO bookshops).

Arising from such a Code, institutions generate "local rules" on Radiation Protection. An example is given in Table 8.3.

Preface

Under the Health and Safety at Work act all institution must provide a safe working place, to ensure that people (including yourself) are not exposed to risks to health and safety.

General rules

1) Coats and other personal belongings including books (except laboratory note books required for immediate use) must not be brought into the Nuclear Laboratory where they may become contaminated.

2) Eating, drinking, storing or preparing food, smoking, or applying cosmetics are all forbidden in any area where radioactive materials are stored or used.

3) Direct contact with radioactive materials must be avoided by wearing the laboratory coats provided and by wearing rubber or disposable plastic gloves. Such protective clothing must not be removed or worn outside the Nuclear Laboratory.

4) Pipetting liquids of any type by mouth, or the performance of any similar operation by mouth suction is strictly forbidden.

5) Work must be carried out under a hood in all cases where gaseous products or radioactive material may be lost by volatilisation, dust dispersion, by splattering or by spraying. Wherever possible work should be performed with closed containers.

6) All radioactive samples must be suitably labelled with the name of the isotope, its activity and date of preparation. Samples must be covered or stoppered.

7) Liquid wastes must not be poured into the spine drain system but placed in suitably labelled containers. Similarly, contaminated glassware must not be washed in the bench sinks.

Table 8.3 Example of local rules on Radiation Protection (This example is based on those operative at the University of the West of England. (Continued)

8) Solid wastes and contaminated articles (corks, gloves, paper wipes etc) must be disposed of in designated containers and must not be placed in ordinary waste receptacles.

9) Laboratory work must be carried out on plastic trays covered with a suitable absorbent material (eg benchkote). At the end of the working period, materials must be left tidily on the trays. All syringe needles must be replaced in their protective covers.

10) Before leaving the Nuclear Laboratory after working with radioactive materials, each person must wash their hands and check them in the hand monitor situated in the exit lobby. Any contamination found upon monitoring must be reported immediately to the lecturer/manager/officer in charge and to the Radiation Protection Supervisor.

11) All spills of radioactive material must be reported to the lecturer/manager/officer in charge immediately. In the event of a spill:

a) any liquid must be blotted up with dry tissues;

b) any dry radioactive materials should be covered with damp tissues and damp conditions used in the subsequent decontamination procedures;

c) attempts must be made to prevent spreading of the spill;

d) the spill area must be isolated, identified as to the nature of the contaminant, and access to the area and any personnel involved must be made immediately, and following decontamination.

12) In the event of a wound that is possibly contaminated with radioactive material:

a) flush the wound in running water;

b) immediately seek the aid of the Radiation Protection Supervisor;

c) an accident report form should be made out following usual local procedures.

13) Any failure to comply with the above Rules must be reported to the Radiation Protection Supervisor.

Additional Rules for working with sealed sources of radiation

1) Never pick up sealed sources with the hands - always use long forceps or reaching tools. Sealed sources must be replaced in their containers after use.

2) Barriers, when appropriate, must be placed where the dose rate is 7.5 uSv/h or above. These must be labelled giving a clear indication of the danger of the radiation. The position of the barriers must be checked by monitoring.

3) Body dosemeters must be obtained from the Radiation Protection Supervisor and worn when working with any source which gives rise to dose rates in excess of 100 m(symbol)Sv/h at or near its surface.

Table 8.3 Example of local rules on Radiation Protection (This example is based on those operative at the University of the West of England.   (Continued)

<table>
<tr><td>4)</td><td>Sources suspected of leaking contamination (eg those damaged or thought to be damaged) must be reported to the Radiation Protection Supervisor for checking. All sources will have been given a Laboratory/Institution identification number and will have been routinely leak-tested every two years.</td></tr>
<tr><td>5)</td><td>Any failure to comply with the above rules must be reported to the Radiation Protection Supervisor.</td></tr>
</table>

Table 8.3 Example of local rules on Radiation Protection. This example is based on those operative at the University of the West of England.

8.6 Disposal of radioactive materials from laboratories.

The disposal of radioactive materials from laboratories and other institutions is, of course, of major importance and is strictly controlled. The incorporation of the provisions of the EC-Directives into national law within EC Member States has led to some national differences, although the standards used are similar in each case.

As an illustration, the Regulations governing the disposal of radioactive substances in the UK, formerly provided for in the Radioactive Substances Act 1960 has been amended by Part V and Schedule 5 of the Environmental Protection Act 1990. Thus it has become part of the Environmental Protection measures operating within the UK. A similar movement of regulations governing the storage, use, accumulation and disposal of radioactive substances into Environmental law is taking place in many States.

Generally such national Regulations establish the following:

- each institution seeking to hold, use and dispose of radioactive materials must register their intention to do so with the appropriate national authority;

- a national authority which sets out the conditions under which a plant, factory or laboratory may operate using radioactive substances. The national authority also sets the discharge limits to air, water or solid discharges and may impose certain conditions on disposal (eg solids may be required to be incinerated);

- an inspectorate that monitors the behaviour (for example the keeping of records) of installations using radioactive substances. Such inspectors are empowered to issue notices to enforce, prohibit and relocate where there is a risk of environmental harm or where the conditions of the authorisations are being contravened.

In the UK, the inspectorate (HM Inspectorate of Pollution) is enabled to charge for their services in respect of considering applications, issuing authorisations, carrying out inspections, and registering companies. Failure to pay fees and charges constitute a breach of registration and authorisation. A growing trend in national legislation is the right of the public to see it the information relating to the application, registration and authorisation of companies or organisations using radioactive materials. Exceptions to the public access to information can be granted for cases which involve commercial secrecy or national security.

Appendix 8.1 Basic information and glossary of terms concerning radiation

8.1 Introduction

Figure A8.1 shows a simple model of an atom. It contains a single nucleus consisting of positively charged protons and neutrons (with no charge) being surrounded by shells of negatively charged electrons (with both inner and outer orbits). Although this model is simplistic it will suffice. We will assume that the reader knows what an isotope is and what the difference is between a stable and an unstable isotope. (A glossary of terms has been placed at the end of this appendix to help you understand any unfamilar terms and we have included many helpful reminders in the next section).

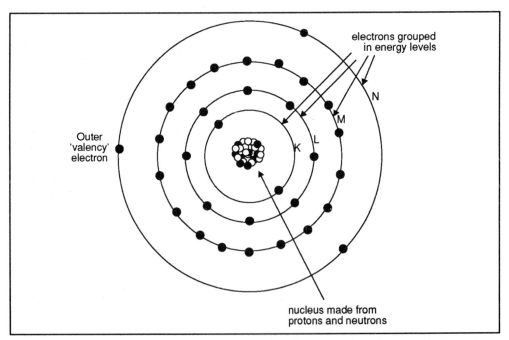

Figure A8.1 Diagrammatic structure (Rutherford-Bohr model) of an atom. The nucleus contains the protons and neutrons around which orbit electrons at fixed energy levels (labelled K, L, M and N). The outermost electrons are the valency electrons.

The process of adding or removing one or more electrons to or from an atom is called ionisation. Particles or rays emitted from radioactive sources can often "collide" and transfer electrons away from or towards atoms and thus cause ionisation; hence they are often referred to as "ionising radiations". Atoms are usually combined with other atoms to form molecules. If ionising particles or radiation should "hit" a molecule it may cause the bonds between atoms to break as electrons are transferred. Living cells contain biological macromolecules (eg DNA) whose structure and function are essential for the correct working of the cell. If these molecules are somehow disrupted by ionising radiation, the direct or indirect effects can be potentially very damaging.

A8.2 Basic information about radiation

For most elements there are isotopes with nuclei that are unstable and therefore undergo radioactive decay. There are two types of radioactive decay depending on the type of emission, which are i) alpha radiation ii) beta radiation. A third type of emission, gamma radiation is commonly associated with i) and ii). All these types of emission are able to cause ionisation and thus potentially damage biological systems. In addition neutrons may cause ionisation through indirect interactions.

8.2.1 Types of decay

Alpha decay:

One way that unstable nuclei with atomic number (Z) greater than 80 can transform into more stable systems is via emmission of alpha particles. This occurs in the transuranic elements such as uranium, thorium and plutonium (see example in Table A8.1). An alpha particle is a helium nucleus consisting of two protons and two neutrons.

Alpha particles move relatively slowly (ca. 5% the speed of light) have a relatively high mass and have a double positive charge. They are therefore very efficient at causing ionisation since their slowness, size and charge ensure that they easily collede and interact with other atoms. However, as the particles transfer their energy they also slow down. Thus, their penetration is very low, particularly when they are travelling through a dense medium.

Beta decay:

β-radiation =
negative
electron

A second way in which unstable nuclei may transform into more stable nuclei is for them to emit electrons or positrons (see examples in Table A8.1). The positrons (positively charged electrons) and negatively charged electrons are called beta particles as they arise from the nucleus and are given the symbols β+ and β- respectively.

Since beta particles do not carry as much charge as alpha particles they are less capable of causing ionisation. However, they are significantly smaller in mass and can move about twelve times faster than alpha particles and therefore are capable of greater penetration.

Gamma emission:

During radioactive decay of some nuclides, the ejection of alpha or beta particles is not sufficient to fully stablise (de-excitate) the nucleus of the daughter nuclides. The remaining energy is emitted as a gamma ray. Gamma rays are a form of electromagnetic radiation (ie photons) which are at the high energy end of the electromagnetic spectrum (see Figure A8.2) similar to X-rays, and are able to eject electrons from atoms following collision and this cause ionisation. Moreover, gamma rays are much more penetrating than alpha or beta particles.

Neutrons:

Neutrons (because they are electrically neutral) cannot directly interasct with electrons. However, they can cause ionisation through indirect actions. This includes i) collision with other nuclei where energy is released from slowing down theneutron ii) capture by nuclei to create a radionuclide which can then decay in the usual way.

1) Alpha-particle emmission

Alpha-particle decay occurs mainly in very heavy nuclides which can attain a stable configuration by emitting an alpha-particle:

$$^{226}_{88}Ra \longrightarrow {}^{222}_{86}Rn + \alpha - \text{particle}$$

$$^{238}_{92}Ra \longrightarrow {}^{234}_{90}Rn + \alpha - \text{particle}$$

An alpha-particle consists of 2 protons plus 2 neutrons bound together. This is the same as a helium nucleus which is why the symbol for a helium nucleus ($^{4}He^{2+}$) is often used to designate an α particle.

2) Beta-particle emission

a) β^- emission

Most of the lighter, neutron excess nuclides, including many of the biologically important elements (eg. H, C, P, S, Ca, Fe, Cl) become more stable through the transformation of one of the neutrons in the nucleus into a proton. During this process an electron is formed (n --> p + β^-).

$$^{14}_{6}C \longrightarrow {}^{14}_{7}N + \beta^-$$

$$^{32}_{15}P \longrightarrow {}^{32}_{16}S + \beta^-$$

b) β^+ emission

For neutron deficient nuclides, protons in the nucleus may transform into neutrons with the loss of a positive charge. The positive charge is in the form of electron (e+ or β+ particle). A positive electron (sometimes called a positron) is the antimatter equivalent of the negative electron.

$$^{22}_{11}Na \longrightarrow {}^{22}_{10}Ne + \beta^+$$

$$^{18}_{9}F \longrightarrow {}^{18}_{8}O + \beta^+$$

3) Gamma radiation

Some unstable nuclei decay in a 2-stage rather than a single stage process. The first stage occurs with alpha or beta particle emission as outlined above. However, with some nuclide, the daughter nucleus still retains too much energy to be stable. Thus, a second stage occurs where the excess energy of the excited nucleus is liberated as gamma rays (ie. production of electromagnetic radiation or photons). The energy possessed by nuclei is quantised, that is they can exist in the ground state and in various discrete excited states. Therefore, the gamma rays that are emitted during the second stage of decay have a characteristic energy corresponding to the difference between the excited state and the ground state, is in turn characteristic of the nucleus from which it arises.

Table A8.1 Types of radiation.   (Continued)

$$^{22}_{9}\mathrm{Na} \longrightarrow {}^{22}_{8}\mathrm{NO} + \beta^+ + \gamma$$

$$^{60}_{27}\mathrm{Co} \longrightarrow {}^{60}_{28}\mathrm{Ni} + \beta^- + \gamma$$

$$^{211}_{27}\mathrm{Co} \longrightarrow {}^{207}_{28}\mathrm{Tl} + \alpha + \gamma$$

4) Other types of decay mechanisms

a) Electron capture

In some neutron-deficient nuclides, a proton can be transformed into a neutron, not by emission of β+ particle (as in beta decay mentioned above), but instead, by capturing an electron from the innermost (K) shell of electrons. The K-shell vacancy is then filled by electons from shells of higher energy levels. The various electronic movements give rise to electromagnetic radiation in the form of X-rays. The wavelength is characteristic of the nuclide.

$$^{54}_{25}\mathrm{Mn} \longrightarrow {}^{54}_{24}\mathrm{Cr} + X-rays$$

b) Isomeric transition

Excited nuclei of some nuclides do not decay instantly; instead they may exist in an excited "metastable" state prior to decaying to the ground state, emitting gamma rays as they decay. The decay is random, but within a finite period of time. Such nuclides which can exist in "high" (excited) or "normal" (ground) energy states, but are otherwise the same are called isomers.

$$^{80}_{25}\mathrm{mBr} \longrightarrow {}^{80}_{35}\mathrm{Br} + \gamma-rays$$

the letter 'm' is used to denote the metastable excited isomeric species.

c) Internal conversion and Auger Electrons

Another way in which excited nuclei can lose their excess energy is to interact with the orbital electrons in the atom (but in a different way to that leading to electron capture mentioned above). This interaction takes the form of i) ejection of an orbital electron followed by ii) electronic rearrangement whereby the electron vacancy is filled by outer electrons. This gives rise to X-rays. This is called internal conversion. In some cases, the X-rays created from internal conversion can themselves interact with the outer shell electrons and eject these. These electrons are called Auger electrons.

Table A8.1 Types of radiation.

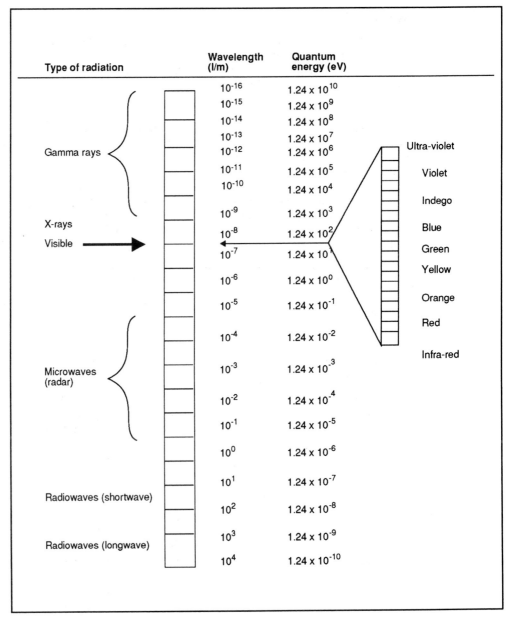

Figure A8.2 Electromagnetic spectrum.

8.2.2 Half-life and activity

Radioactive decay:

The term "half-life" is often used to give a measure of radioactive decay. This is the time it takes for the activity of a sample to fall to half its initial value and can vary enormously from milliseconds to millions of years depending on the radionuclide in question. Table 8.2 gives examples of radionuclides, their half-lives and their principal emissions.

Radioactivity:

Radioactivity can be measured as disintegrations per second. The standard unit for activity is one disintegration per second and is given the name Becquerel (Bq). However, the Becquerel is often too small to be useful and so the kiloBecquerel (kBq) and the megaBecquerel (MBq) are more often found. These are 10^3 and 10^6 Bacquerels respectively.

The half-life of a radionuclide is characteristic of that nuclide and independent of the mass or form that is present. This is, of course, not so for the radioactivity. For example, 1 gram of radium has an activity of 3.7×10^{10} Bq, whilst 10 gram has an activity of 3.7×10^{11} Bq. Both have the same half-life.

In practice, activity is measured by counting the number of alpha particles, beta particles or gamma rays emitted per unit time. The rate at which this dimishes over time can be used to calculate the half-life. Alternatively, if the half-life is known, the amount of radionuclide in a sample can be calculated from measuring its activity.

Element	Atomic weight	Radionuclide	Half-life	Principal emissions
hydrogen	1,008	^{3}H	12.3 years	β-
carbon	12.01	^{14}C	5,730 years	β-
sodium	22.99	^{22}Na	2.6 years	β+
phosphorous	30.97	^{32}P	14.3 days	β-
sulphur	32.06	^{35}S	87.4 days	β-
chlorine	35.45	^{36}Cl	300,000 years	β-
potassium	39.10	^{42}K	12.4 hours	β-
		^{40}K	1,260,000,000 years	β-; γ
calcium	40.08	^{45}Ca	165 days	β-
strontium	87.62	^{90}Sr	28.1 years	β-
iodine	126.91	^{131}	8.0 days	β-, γ
radon	222.02	^{220}Rn	55 secs	α
		^{222}Rn	92 hours	α
radium	226.02	^{226}Ra	1602 years	α, γ
uranium	238.03	^{233}U	162,000 years	α γ
		^{235}U	710,000,000 years	α, γ
		^{238}U	4,500,000,000 years	α

Table A8.2 Examples of radionuclides, their half-lives and their principal emissions.

A8.3 Effects of radiation on cells

Animals, plants and micro-organisms consist of cells each of which contains nucleic acid surrounded by the cytoplasm and contained within the cytoplasmic membrane. In higher plants and animals (and fungi) the DNA within the cell is contained within a nucleus on chromosomes which are made of DNA bound with proteins. DNA contains the coded information (blueprint) required to direct the development and function of the cell. It does this through the expression of genes. A gene is a segment of DNA which contains the information to make a single specific protein. Thousands of different genes are required in order to code for a whole cell.

In addition to coding for proteins, DNA is the vector by which genetic information is passed on to new cells. Thus, in cell multiplication (mitosis) DNA replicates prior to the cell dividing into two.

mutation

Radiation can induce changes in the genetic material of cells. Changes in a single gene (gene mutation), or in the structure or the number of chromosomes (chromosomal mutation) can occur which may kill the cell or give the cell varieant properties which were not present in the original cells. Such mutations can induce the tranformation of a normal cell into a cancerous cell.

generation of toxic free radicals

There are two ways in which ionising radiation can cause damage to DNA. Firstly, alpha, beta or neutron particles can directly "hit" the DNA molecule and cause breakage or aberrant "cross-linking". Secondly and more commonly, radiation (all types) can cause ionisation of water molecules within the cytoplasm and induce the formation of toxic free radicals which in turn can damage the DNA Radiation can also damage cell membranes and distrupt cell transport functions.

Cells vary enormously in their sensitivity to radiation. Micro-organisms are generally more tolerant towards ionising radiation than are higher organisms. In higher organisms, tissues which have a high cell turnover more rapidly show the effects of radiation than those in which the cells are growing very slowly.

non-stochastic effects = magnitued of effect is proportional to dose

Somatic effects (ie damage to the body of the exposed individual ulcers) are well documented and range from skin burns, ulscers, dermatitis, diarrhoea, hair loss to anaemia and other blood changes. Acute radiation effects in humans (eg skin burns, radiation sickness, cataract of the eye lens, effects on bone marrow stem-cell production) are all examples where the radiation dose has to be above a threshold level. The effect is said to be "non-stochastic" or deterministic. The damage results from cells being killed. If only a few cells are killed in a tissue or organ, the tissue or organ can still function and new cells can replace the ones that have died (regeneration). Only if the radiation is above a certain threshold, when a significant proportion of cells are killed in, the tissue will damage become obvious. Longer term effects can include various forms of cancer. In cancer, the radiation dose may not be high enough to csuse significant lethal damage to cells in a tissue but may cause genetic damage in some. The mutated cells may continue to divide and give rise to large numbers of damaged progeny. Mutation rates increase proportionately with dose of radiation there is no threshold for the effect to occur (ie no minimal dose) and the probability of its occurrance increases with dose: thus the effects are said to be "stochastic".

strochastic effects = probability of effect is proportional to dose

Hereditary effects (sometimes called "genetic" effects) are those that show themselves in the descendants of the exposed individual. Studies of irradiated insects (*Drosophila*) and small mammals have shown that radiation-induced gene and chromosomal mutations can cause a wide range fo abnormalities in the offspring. The evidence in humans is less marked. Nevertheless it is prudent to cosnider hereditary effects as stochastic and to accept that there is no threshold dosw below which it could not occur at all. For radiation doses above 1 Siervert (for explanation of this term, see below), non-stochastic effects dominate whereas below this stochastic effects dominate.

A8.4 Radiation dosage

Gray = unit of absorbed dose

Radiation, like pharmacological drugs, produces biological effects. Thus, for radiation the term "dose" has been adopted by analogy with medicines. However, the analogy is false since drugs are measured by their biological effect, whilst radiation is defined in purely physical terms as the amount of energy that is deposited in a unit mass of tissue. This quantity is called the absorbed dose, and the SI unit for this is the Gray (symbol Gy) which is equal to an energy deposited by any type of radiation of 1 joule per kilogram of tissue.

quality factor (Q)

However, the absorbed dose alone is not a very good predictor of the severity of the effects of radiation since it is the form in which the energy is delivered rather than the total amount of energy that is important. It is in recognition of this that the International Commission on Radiological Protection (ICRP) introduced a quality factor (Q or QF for short) to take account of the different damaging potential of different forms of radiation. These are:

For X-rays, gamma rays and beta particles QF = 1

For neutrons (slow, medium or fast)   QF = 5, 10 or 20

For alpha particles   QF = 20

The dose equivalent can now be calculated as the absorbed dose x QF

Sievert = unit of dose equivalent

The unit of dose equivalent is the Sievert (symbol Sv). When attempting to assess a radiation hazard it is important to know the rate at which radiation is being absorbed. Rates are measured as dose per unit time; thus units are Grays per hour or Sieverts per hour. However, the Gray and the Sievert are usually too large a unit for practical purposes, therefore milliGrays or milliSieverts per hour or micro Grays or micro Sieverts per hour are used instead.

A8.5 Glossary of terms

Alpha particle: A relatively stable association of 2 protons and 2 neutrons which can be emitted during radioactive decay. It also constitutes the nucleus of the helium atom

Atom: The smallest unit of a chemical element.

Atomic number (of an element): The number of protons in the atomic nucleus. It is equal to the number of electrons in a neutral atom.

Atomic mass number: The total number of protons and neutrons in the nucleus.

Beta particle: A fast electron or positron emitted during the radioactive decay of a nucleus.

Electron orbital or energy levels: Electrons in atoms can be assigned to different orbitals. The K orbital is closest to the nucleus and has two electrons, the next shell (L) has 8 electrons and so on. Each element is composed of atoms with the same characteristic orbital structure. The chemistry of the elements is strongly influenced by the nature and number of electrons in the outer orbitals.

Element: A substance which cannot be broken down into simpler constituents by chemical means because all its atoms have the same atomic number. The naturally occurring elements range from hydrogen (atomic number 1) to uranium (atomic number 92). Elements with higher atomic numbers (transuranic) have been produced artificially.

Gamma rays: Electromagnetic radiation emitted in the form of photons during radioactive decay. Gamma rays are at the high energy end of the electromagnetic spectrum and (in contrast to visible light) are able to eject electrons from atoms following collision. They are thus said to be ionising.

Half-life: The time which must elapse for half the activity in a large sample to radioactively decay.

Ion: An atom or molecule which is electrically charged because of an excess or deficiency of electrons. If the atom loses electrons it becomes a positively charged cation; if it gains electrons it becomes a negatively charged anion.

Isotopes: Atoms of the same chemical element, having the same atomic number, but differing from each other in respect of the number of neutrons in the nucleus eg deuterium is an isotope of hydrogen. Most isotopes found in nature are stable but some are unstable, and their nuclei undergo radioactive decay.

Linear Energy Transfer (LET): This is the energy loss per unit of distance travelled. Radiation can be classified into high LET (alpha particles and neutrons) and low LET (beta particles and gamma rays).

Neutron: An electrically neutral elementary particle with a mass slightly larger than a proton. The nucleus of an atom consists of protons and neutrons. In isolation, neutrons are unstable and undergo radioactive decay with a half-life of about twelve minutes into a proton, an electron and a neutrino.

Nuclear fission: The splitting of an atomic nucleus of high atomic number, usually by free neutrons during a chain reaction. The process can be spontaneous or artificial. The nucleus is split into two smaller nuclei and usually a number of other particles including more neutrons. For heavy elements the total mass of the fission fragments is less than that of the original nucleus plus neutrons. The difference is accounted for by the energy released.

Nuclear fusion: The production of an atomic nucleus by the union of two lighter nuclei in a nuclear reaction. Being positively charged, the two original nuclei repel one another, and considerable kinetic energy (ie heat) is required to get the reaction to proceed. However, the energy released can be much greater since the resulting nucleus is less massive than its constituents, the balance being converted to energy according to the mass-energy equation of Einstein.

Nuclear medicine: A branch of medicine that deals with the administration of radioactive isotopes for diagnosis and treatment.

Nucleus: The central core containing most of the mass of an atom. With the sole exception of hydrogen (which consists of a single proton), the nucleus is composed of protons and neutrons held together by strong interactions. Nuclei of the transuranic elements are all unstable because the number of protons is sufficient for their long-range electrostatic repulsion to overcome the short-range strong interactions.

Proton: A positively charged elementary particle about 2000 times heavier than the electron. Protons are stable and found together with neutrons in the nucleus of atoms.

Radiology: The medical application of X-rays.

Radioisotope or radionuclide: Any isotope of an element which is radioactive. Some can occur naturally but others may be produced artificially by neutron irradiation or bombardment with light nuclei. They are widely used in diverse fields including medicine (eg for radiotherapy), in geology and archaeology (eg radiocarbon dating) as well as in biology (eg radioactive tracers to provide information on pathways, reactions, binding, diffusion and transport).

Radiotherapy: The use of ionising radiation in the treatment of cancers.

Radon daughters: The short-lived decay products of radon-222 which are polonium-218, lead-214, bismuth-214 and polonium-214.

Tritium: Isotope of hydrogen containing two neutrons and one proton.

Patents and biotechnology

Patents and biotechnology

9.1 Consumer warning

What follows is an attempt to give readers an understanding of the basic principles of the relevant laws and how they apply to biotechnology. It is not a do-it-yourself guide. Even if the text were absolutely clear and infallible, the law changes. Moreover, though there is a common core of intellectual property law whose principles apply worldwide, the devil is in the detail. The details differ from country to country - and they change frequently. If the matter is of any importance, do not assume that you know the answer - **take good local professional advice**. Our present aim is primarily to teach you enough to alert you to situations in which you need advice.

9.2 Options for protection of intellectual property

In the biotechnology field, there are four options for protecting intellectual property:

- patents;
- confidentiality/trade secrets;
- plant variety rights;
- others.

These are given in general order of importance. They will be considered in more detail, in the same order.

9.3 Patents

9.3.1 What are patents for?

There are two basic purposes:

- to reward inventors;
- to promote technical development.

The first purpose is often to the fore in the minds of constitutional theorists. No doubt, as a matter of equity, inventors should be rewarded for the fruits of their labours. They deserve well of humanity (or most of them do). But whether any system (let alone the patent system) can do this equitably is another matter. Few inventors (even if successful) get rich. Nevertheless the patent system does provide a way for the small person to extract recognition and reward from an idea which otherwise would be free for all to use as soon as it became public.

From the point of view of society, a more important function is the promotion of technical progress. The patent system does this in two ways:

- it encourages investment in new technology;

- it encourages the publication of new technology.

deterrents in investment   Commercially, the way of the innovator is hard. If you are selling something new, you do not know whether there is a market for it. Often there is not - market research is not an exact science. If there is a market, there will be competitors in it as soon as your success becomes known. The competitors will benefit from your failures as well as your success - the second product into a market is often much better technically than the first. Some of your initial investment your competitors will not have to repeat - in consequence they will often be able to sell cheaper than you. You will have to drop your prices to meet theirs, and you will not be able to recover your start-up costs. The fear that this will happen can be a strong deterrent to investing in innovation - private companies cannot invest in projects where they foresee that they will lose money.

Without a patent system, entrepreneurs will have to keep secret as much as possible of the 'know how' they develop. The patent system counteracts this. In order to obtain their rights, entrepreneurs have to disclose their inventions to the public - how to make and use them, as well as what they are. This disclosure in itself encourages the development of new technology.

9.3.2 What is a patent?

A patent is the right:

- to exclude others;

- for a limited period;

- from commercial use;

- of your new invention.

Each of these aspects of the right needs further consideration.

"to exclude others"

The patent right is a private civil right, usually enforced by a private legal action against the infringer. Generally, patent infringement is not a criminal act: it is the responsibility of the patent owner (and no one else) to prevent infringement.

Note particularly that the grant of the patent **gives no rights to use the invention**. Such use may not be possible for a variety of reasons, both practical and legal. Practical difficulties may be - the invention may not work effectively: it may be too expensive to manufacture; there may be no market for it.

Legal difficulties may include the invention being incompatible with existing regulations (for example, for reasons of safety). Sometimes such regulations have been framed with insufficient foresight and can inhibit genuinely useful advance. Much the most common cause of legal difficulties is that the invention cannot be used without infringing an already existing patent. In such cases, you must first negotiate a licence - if the earlier patentee is willing to do this.

"for a limited period"

A patent is not a permanent monopoly. Most countries give a period of protection of around 15 - 20 years. In Europe, the period is generally 20 years from application. In USA the period of protection is 17 years from grant (this may result in both US and European patents expiring at about the same time, since US patent prosecution often takes 2-3 years). During its life the patent is subject to renewal fees - if these are not paid, the patent lapses. After a patent expires, all further use of it is free to all (at least to the extent that such use does not infringe other patents still in force).

"from commercial use"

A patent is directed to promoting commercial exploitation. The patentee gets rights only to control such exploitation-NOT to prevent other use. Most countries recognise a "research exemption" which allows the use of patented inventions for research proposes. Note however that the exemption is for research on the invention, rather than research with the invention. People are not entitled to construct their own research instruments to avoid payment of royalty. They are however entitled to experiment with them to try to improve them or to make alternatives. Private use is also not covered. If you buy a patented watch abroad, you can import it for your own use - but not for sale.

9.3.3 What conditions must a patent meet?

To be patentable, an invention must be:

- new;

- inventive;

- industrially applicable;

- owned by the applicant;

- for permitted subject-matter;

- repeatable from description;

- drafted with claims commensurate with the invention.

These requirements are reviewed in more detail below.

"New"

All patent laws require inventions to be new. They differ somewhat in what they take into account in considering novelty. In principle, a disclosure of how to make the invention made known before the patent filing date invalidates the patent. In European law, such disclosure may be made anywhere in the world, and may be in writing, in a diagram, by use, or even orally. No exception is made for disclosure by the inventor - the inventor must file before disclosing or the patent will be invalid. Unfortunately this is often overlooked and many patents are invalid in consequence. In other countries (for example New Zealand until very recently) publication will not invalidate unless in the local country. In USA, on the other hand, prior disclosures must be in a printed document, or by way of use in the USA: and disclosures by the inventor do not invalidate unless made more than one year before the US filing date. Earlier filed patent applications are also taken into account in considering what is new.

"Inventive"

It is not sufficient for the invention to be new. It must also be unobvious, or "inventive". What constitutes invention is not easy to define - it is more often defined by what it is not. A development is not inventive if it is obvious. It is obvious if it is "lying in the way" (Latin *ob via*) - if it is what anyone who thought about the matter (and was aware of what had been published) would naturally do the same when faced with the problem the invention solves. Thus it can be persuasive of inventiveness if the applicant can show that others had been faced with the problem and had not been able to come up with the new solution. Deciding whether a new development is or is not obvious is one of the most difficult tasks for a Patent Office or a Court, and such decisions are difficult to forecast.

Generally, what is taken into account in considering whether the invention is obvious is the same as for novelty - what has been published or otherwise become known. There is, however, a difference between US and European practice here. In Europe, a prior filed European patent application is considered for novelty only and not for obviousness. This often allows a second applicant to get a patent for simple and obvious variants of an earlier unpublished invention. In US, it is considered more important not to grant more than one patent for the same invention. A prior patent can be used to make the invention obvious, even if not published at the later patent's filing date. However a patent applicant can overcome an earlier published document (provided it is published less than one year before his US filing date) by showing that he made the invention earlier (see discussion of Interferences, later).

"Owned by the applicant"

Inventions belong in the first place to their authors. If the inventor is employed to make inventions, those inventions belong to the employer. In Europe, an applicant for a patent must name the inventor or inventors, and show the chain of title (an assignment - either direct or indirect - from the inventor to the applicant). In USA, patent applications can only be made by the inventors - but employed inventors then usually assign such applications promptly to their employers. An applicant who appropriates the invention of another without consent can have his patent transferred to the correct owner.

"For permitted subject-matter"

Most countries exclude claims for certain types of subject-mater. Typically, as countries become more developed, they permit a wider range of inventions to be claimed. Thus, up till 1950 Britain allowed claims only for chemical processes and not for chemical products. In West Germany (as it then was) and Japan, such claims were not permitted until the 1960's. Many developing countries still have restrictions on claiming both chemical compounds and pharmaceuticals (for example India, Brazil): indeed, Italy only began to allow claims to pharmaceuticals within the last few years. Other countries have more specialised exclusions - for example, Thailand does not allow the patenting of agricultural machinery. Several countries have restrictions on the type of protection obtainable for biotechnology inventions - of which more later.

"Repeatable from description"

It is a vital part of the bargain that the patentee makes with the public that he discloses fully how to carry out his invention. This used to be done by training apprentices - indeed, the original term of a UK patent was set at 14 years because it gave time to train two generations of apprentices in the operation of the invention. Nowadays this disclosure is carried out in principle entirely by the written description ("specification") that the patent applicant files with his patent application. This description is addressed to people skilled in the field that the invention relates to - "the art", as it is termed - hence the phrase "state of the art". The addressee is expected to know all that anyone of average skill in the art would know: if more information is required, the patent applicant must supply it, or give a reference to it if it is already published. If suitable information to carry out the invention is not given, the patent is invalid for "insufficiency".

One exception to the rule of written description, of substantial importance to biotechnologists, has recently evolved. Living material cannot generally be described in a way that allows it to be produced from readily available materials. Numbers of inventions however use such living materials as essential starting points. For example, particular strains of micro-organisms produce novel antibiotics: the only way of making such antibiotics may be by fermentation of such micro-organisms. To overcome this difficulty, it is possible in a number of countries to deposit a sample of such a micro-organism in a public depositary. When the patent specification is published, samples of the micro-organism can be made available to the public on request (under certain conditions). This deposit supplements a full written description of the micro-organism and ensures that the invention is fully made available to the public.

This facility to deposit is often made use of by biotechnological patent applicants, to supplement written descriptions that might otherwise be incomplete. But there is no requirement to deposit any organism or material if it can be obtained from what is already available. For example, if the invention employs a particular DNA sequence, it would not be necessary to deposit that sequence if it was short enough to be synthesised, or if instructions could be given for making a cDNA library from readily available materials and isolating it from that.

"Be drafted with claims commensurate with the invention"

This is a rather less specific requirement, and one which does not necessarily appear formally in patent laws. It is nonetheless real and important. It is in part the question of the way in which claims are framed. A claim should define a solution to a technical problem, not simply restate it. This is well recognised in certain technical areas, for example in the mechanical and pharmaceutical arts. In no circumstances would a patent examiner allow a claim to an engine defined solely by its power/weight ratio, or a claim to a pharmaceutical defined solely by its therapeutic activity. There is however a tendency in the biotechnology area to define inventions primarily in terms of obviously desirable properties - a new strain of wheat, for example, might be defined solely in terms of its resistance to insects. This arises from a commendable wish to do justice to meritorious inventions, but fails to take into account of the position of those no less worthy which achieve the same or even better results by entirely different means. Too broad claiming is not fair to the public and holds up the exploitation of the technology.

9.3.4 Other requirements

The two most frequently met with are **utility** and **disclosure of the best mode.**

Utility

invention
should be
useful

It used to be a requirement of British law that the invention be useful. The standard of utility was broad but not high. That is to say, the invention did not have to have commercial utility (in the sense that it could be made economically, or that there must be a market for it). But it was necessary that the invention should do what the patentee claimed it would do - that it would carry out what the patentee promised. And this promise had to be fulfilled not merely by whatever the patentee described, but also by whatever fell within the scope of his claims. This resulted in promises in patent specifications being drafted with some care! Current European law does not have any requirement for utility on the theory that it is not relevant: any patent infringement will certainly have utility, or it would not take place. But this fails to take into account that the utility claimed or demonstrated by the patentee might be different from or very much less than that of the infringer. The requirement for utility is now shaded into that for industrial applicability - completely useless inventions (eg perpetual motion machines) are held not to be industrially applicable. USA, however, still retains a utility requirement.

Disclosure of the best mode

This requirement also used to form part of British law, but does no longer. It is still very important, being a requirement in USA and Australia. The bargain between the patentee and the public should require good faith on the patentee's part: he should not disclose an inferior form of his invention and keep the best information to himself. This requirement serves to keep the patentee honest. In theory it might be possible to evade it by not filing in USA or Australia, but most important inventions will require protection in one of these countries. Different disclosures in different countries would simply highlight the information it was desired to conceal.

9.4 What can be claimed?

crucial nature
of claims

The claims of a patent are crucial to the protection it gives: they are to define exactly what it is others cannot do (but see also the next section). Equally, it is what falls within the claims that has to meet the criteria of novelty and inventiveness.

Claims are generally to **processes** or **things.** In some countries (and under EPO practice) claims to "uses" are allowed, but these are really a type of process claim. In a mechanical case, for example, the applicant will usually claim a machine, sometimes a process of making it, or using it. In a chemical case, the applicant may claim a new chemical compound or composition, or an improved process (or method) of making a chemical compound. A new use for a known compound may be claimed.

While claims are obliged to define the invention clearly, they are not obliged to be specific, or indeed limited to the specific working examples that the patentee has given. Many claims use functional definitions, at least in part. For example a new design of waterproof garment may be defined as being made of "flexible impermeable material". The patentee will not normally define precisely what materials can be used (nor even exactly what standards of flexibility and impermeability are implied). This is reasonable, because the invention is in the design of the garment, not the material to be used - examples of suitable materials will usually be given in the specification, but the person skilled in the art may be expected to know what is suitable. It might not be necessary to include any definition of the material in the claim, but this will not be so if the invention consists in applying, to waterproof garments, features that are already known in sweatshirt (which are permeable) or suits of armour (not flexible). This type of functional definition is appropriate in this type of situation, but not where the feature of the claim being defined is the crucial point of novelty. For example, the inventor of non-stick frying pans should not be permitted to claim "A frying pan coated with a heat-resistant non-stick plastic coating", because that is a claim to the problem, not the solution to the problem. The solution is the specific polymer that the inventor has found that will meet these requirements - or the way of bonding the plastic to the metal base.

What can be claimed has developed over the years. The original British concept (now superseded) limited claims to "vendible products" and processes of making them. The International Convention on Patents refers to the patentability of inventions useful in "any kind of industry, including agriculture".

9.4.1 Claims in biotechnology inventions

In the light of this, we can review what protection is available for a biotechnology invention. Obviously this depends on the nature of the invention. Claims are allowed for both processes and products. Examples of products are:

- DNA sequences;

- DNA constructs;

- plasmids;

- proteins;

- organisms.

What are the limitations on such claims?

patenting of living organisms

Firstly, there are problems (real and perceived) with patenting living organisms. Can life forms be patented? Until recently, claims to organisms have been rare, in nearly all countries. The USA has a doctrine that "products of nature" are unpatentable, and all living organisms were originally seen as "products of nature". In other countries, a difficulty in patenting organisms was the impossibility of giving a description which the reader could repeat. Processes of breeding plants and animals were not specifically reproducible, and in any case generally needed access to specific parents. Nevertheless, claims to living organisms were granted occasionally - Pasteur was granted patents to yeasts in the last century in both the UK and the USA.

case histories

In the USA, the question was definitively answered by a Supreme Court, Diamond v Chakrobarty 44US.303 (1980). Micro-organisms that had been genetically modified so as to clear up oilspills were held to be patentable: the Court held that "everything under the sun produced by the hand of man" should in principle be patentable. This decision on micro-organisms has since been held to justify patenting plants (*Ex porte* Hibberd - maize containing higher than normal quantities of a particular amino acid nutrient) oysters, and most recently a genetically modified mouse, for use in cancer research (the so-called "Harvard Mouse", the patent being in the names of the President and Fellows of Harvard College - as owners, rather than inventors).

role of the
European
Patent
Convention
and the
European
Patent Office

In Europe, the position is at present controlled by the European Patent Convention (EPC). Section 53 of this convention specifically excludes "plant or animal varieties or essentially biological processes for production of plants or animals; this provision does not apply to microbiological processes or the products thereof". The Board of Appeal of the European Patent Office has ruled that exceptions to patentability are to be construed narrowly (and hence, that exceptions to exceptions are to be construed broadly. "Plant variety" has been held to have a specific meaning (a variety of the type that can be protected by a plant variety right - see below), and patents have been granted to plants (specifically genetically modified plants) that would not qualify as "varieties". The EPO has also allowed the patent application to the "Harvard mouse". Both plant and animal patents are controversial in Europe, and oppositions to both types of patent have been filed by environmental and animal protection groups. These groups consider such patents to be contrary to morality, and hence forbidden by another article of the EPC. The belief of the biotechnology industry (possibly based on wishing it to be so) is that patents of this type will be upheld in Europe, as they are in USA.

Any biotechnology patent has to meet the normal standards of patentability as set out in Section 9.3.3 above. There is however a question of how to apply these standards, particularly novelty, to inventions which consist in isolating useful products from nature. It is generally considered that an isolated natural product is not automatically unpatentable because it has always existed in nature. It will not be patentable to isolate a product that is previously known to exist and have useful properties. Such isolation would be obvious - unless, perhaps, there were special difficulties about the process of isolation. Furthermore, the claims must be drafted in such a way as to distinguish the isolated product from the product in its natural state. Claims to natural proteins, for example, usually specify some degree of purity: proteins made by recombinant micro-organisms may be defined as free from glycosylation (natural plant and animal proteins being normally glycosylated).

How does this apply to patenting DNA, and in particular genes? There is no problem in principle in patenting newly synthesised synthetic DNA sequences. But what about genes isolated from nature?

A distinction is often made between "inventions" and "discoveries": the former are said to be patentable, the latter not. This doctrine is argued to prevent the patenting of genes, as these are discovered rather than invented. But the doctrine is too simplistic, and hence not useful. A discovery in science is usually of some new effect, or of a mechanism which underpins or explains some existing effect. Discovery of a new useful effect normally implies a new process or product which can be patented: elucidation of a known process often suggests new means of putting it into effect. For example, discovering that a known compound kills weeds leads to new herbicidal processes and compositions which can be patented. Where this is not so (as for example, where the mechanism of a known effect is explained (Brownian motion, say, or the theory of gravitation) the new knowledge is (in patent terms) "a mere discovery" because it does not lead to any new process or thing that can be patented. But this definition is simply circular and leads nowhere.

Given that a gene already exists in nature, its isolation and sequencing is a discovery. Is it a "mere discovery"? If the existence and function of the gene are known, it may be. This question is not really settled either way, and perhaps there is no general answer, only a specific answer in the circumstances of each case. Knowledge of the new sequence does make it possible to do things (such as producing the protein in new organisms, or inhibiting its production in organisms that normally produce it) that have not been done before. But novelty is not enough - inventiveness is also required. In any particular case, the patent applicant may make two arguments:

- the gene was difficult to identify or isolate - as evidenced, perhaps, by others having tried and failed;

- though there was no particular difficulty in isolating the gene, there was no particular reason to, either: the patentee has discovered some advantage in the use of the gene that had not previously been apparent, but for which, no-one would have concerned themselves with isolating the gene.

Either argument may fail, or succeed, on the particular facts of the case. In principle, there should be no rule that the first to sequence a particular gene is entitled to a patent for it, because there is no guarantee that this will constitute an invention. Nevertheless, most patentees in this situation get their patents allowed. Only one so far has been found invalid: this was the Genentech patent on the TPA protein and DNA sequences coding for it, which was struck down by the British Court of Appeal. Here TPA was a recognised important commercial target, three or four research teams were working on the problem of isolating the gene, and a partial sequence was already available. Genentech were the first to obtain the full sequence, as a result of much careful teamwork, but without (in the Court of Appeal's view) doing anything that was not in principle suggested by the prior art. In these circumstances, the claims to the sequence were found unpatentable. As an indication of how uncertain this conclusion is, however, both the US Court and the European Patent Office have upheld the Genentech patent.

In many cases, there will be an invention to be claimed. Then the DNA sequence should be distinguished from the sequence found in nature. Some argue that this is not necessary, but it is still strongly recommended. Otherwise the question will arise as to whether the patent can be asserted against the gene in natural pre-existing organisms. A court will certainly find that the answer to this question is No. To reach such a conclusion, the Court will place its own limitation on the meaning of the claim, which may be much more severe than the patentee would like - the Court might say, for example, that the patentee intended only to claim isolated DNA sequences, not sequences incorporated in a genome. The patentee should therefore make his own distinction, and not leave it to the Court.

A common way of making the distinction is to recite that the DNA is "recombinant": other possibilities are to claim it in an alien environment, such as a vector; or as cDNA, which does not occur in nature.

9.5 Patent infringement

The effects of a patent are national and territorial. Broadly, the claim defines the scope of the protection - but this simple principle requires careful qualification. In general, a patent is infringed:

- by making, possession or use of what the claim specifies within the jurisdiction while the patent is in force;

- by inducing others to infringe.

Note however that infringement is a matter of substance. Immaterial limitations in a claim are liable to be disregarded. Indeed, increasingly the claims are seen as defining the scope of the invention, rather than the nature of the infringing acts. A court trying to find the meaning of a claim has to find a fair balance between the patentee and the public. Formerly it was considered the duty of the patentee to draft his claim with care so as to cover the invention fully and distinguish what was his from what the public remained free to do. "What is not claimed is disclaimed" is the classic statement of this viewpoint (by Lord Russell of Killowen, in the British case of EMI v. Lissen). But today British Courts are more influenced by European ideas. The protocol to the EPC provides that claims are to be interpreted in a manner that "combines a fair protection for the patentee with reasonable degree of certainty for third parties". More emphasis seems to be placed on the first term of the equation, however: the result is that it is much more difficult than formerly to be confident of the limits of a patent claim and to be certain that one is free from all possibility of accusation of infringement.

"what is not claimed in disclaimed"

A basic principle of patent law should be that the patentee is only entitled to monopolise what is new and not obvious. In the British courts a special defence used to be based on this: the "Gillette Defence", which would plead that at the date of the patent, the alleged infringement was not novel or was obvious. If this contention was upheld, it was irrelevant to the defendant whether the patent was invalid or not infringed, or both. But nowadays it is uncertain whether such a defence would necessarily succeed. In Germany, for example infringement and validity of patents are judged in different courts: this enables the patentee to put forward a narrow meaning of the claim when validity is in question, and a broad meaning when infringement is being considered. This is hardly fair.

As well as "infringement in substance", claims may be broader than their apparent scope in other ways. Process claims cover products of the process. In Europe, this cover is limited to the "direct product" of the process. Thus, a claim to a new method of manufacturing a plastic would extend to bulk material (powder or granules) prepared by the method, but not to articles (film, fibre, washing-up bowls) made from such material. Where manufacture is carried out within the country where the patent is in force, the extra protection is of little advantage to the patentee: the advantage is that imports can be prevented.

9.5.1 Exceptions to infringement

Patents are concerned with commercial exploitation of inventions. Accordingly there are a number of defences which are normally available to a charge of infringing use. Use does not infringe if it is:

- private and non-commercial;

- experimenting on the invention;

- to fulfil a medical prescription;

- on certain vehicles in transit.

"Private and non-commercial" Under this provision anyone is free to emply an invention for personal use. A person could import and wear a patented watch: perhaps even several watches, if they were presents for family members. A single personal computer, for use in a family business, would in principle be infringement.

"Experimenting on the invention" is a narrower exception than is sometimes realised. There is no general "research exemption" as is sometimes said: this would be unfair to inventors of scientific instruments. There is a specific exemption for experimenting with an invention, to see if it works, to see how it might be improved, and (probably) to see if the patent on it can be avoided. As well as this, some research may be "private and non-commercial" - and in any case it is unlikely to be worth the patentee's while to challenge research use, even where he knows about it. However, experiments to develop an invention (with a view to marketing it on the day the patent expires) have been held to infringe, and are prohibited by injunction, in several countries.

Another very important class of non-infringing acts are those done with the patentee's permission - licensed acts. The patentee's licence may be express or implied. An express licence is a common way of exploiting a patent - the patentee grants a licence to one or more licensees to manufacture and sell the invention in return for payment of a certain royalty - 5% of the selling price, say. In such a case, the licensee's acts clearly are legal, and the patentee may not sue to prevent them: that would be obviously unfair. Equally, those who buy the invention from a licensee, or the patentee, may not be sued either. They are "implied licensees": obviously when they buy the patented article from the patentee (or the licensee) they expect to be able to use and sell it without further interference. Note however that it is not obligatory for a patentee to sell on these terms - more limited rights may be imposed, if this is made clear at the time of sale. Another term for this implied licence is "exhaustion". The patentee's right is said to be "exhausted" by the sale of the product, and thereafter it cannot be asserted.

9.5.2 Biotechnology infringements

Two particular questions arise.

A)   What activities are permitted by the "research exemption"?

Many basic biotechnology processes are patented: for example, the Cohen and Boyer patent, in USA, covers the basic principles of genetic engineering. Is it lawful, in USA, to cut and splice genes without the licence of Stanford University, owners of the Cohen and Boyer patent? In principle No - though Stanford might not sue University staff carrying out an experiment for academic purposes. Even if infringement was found (which would not be certain, as US law talks of an exception for 'philosophic enquiry') damages would be minimal, and quite likely no injunction would be granted. A commercial firm would not get let off so easily - nevertheless, even in the case of commercial research, laboratory work is unlikely to be contentious until it approaches the stage of commercial development. Large-scale trials preparatory to commercial launch have been held to be patent infringement both in USA and Europe - though clinical trials of a new pharmaceutical to satisfy regulatory requirements have been exempted in USA by a special statute. As a guideline, whatever the formal legal position, experimental laboratory work is unlikely to be objected to by the patentee unless it is depriving him of a market for something he is selling (for example, a diagnostic kit).

B)   What activities are covered by implied license or "exhaustion"?

The purchaser of a patented article (from a patentee) normally receives the full right to use that article for its intended purpose. This includes the right to repair it, but not to replace it (case law has drawn this line with some care). It emphatically does not include the right to multiply the article. What then is the position where (as with seed sold to a farmer) the intended purpose of the article is to reproduce itself? There have been no court decisions on this point yet, and not unnaturally the point is hotly disputed between biotechnologists and farmers. The best answer seems to be that someone who sells patented seed to a farmer clearly must expect it to be planted and harvested, so this use is licensed, as is sale of the harvest for consumption - but that subsequent replanting is not licensed and would count as infringement. There is no guarantee that this is the right answer: and in any case it is hardly apt to cover the case of someone who sells viable micro-organisms, for example brewer's yeast. The advice to such a patentee must be not to sell his product, but only to license its use.

9.6 Patent procedure

9.6.1 Application

When an invention suitable for patent protection has been identified, the first step is to file a priority patent application. This gives priority for a period of one year in countries that are members of the International Patents Convention (see below). Before filing, check:

- who are the inventors?

- do they agree to the application being filed?

- are there relevant security implications?

Inventorship

As noted above, an invention in the first place belongs to its inventors. Many inventors are employed to invent, and are obliged by the terms of their employment (or in the UK, by Section 38 of the Patents Act 1977) to assign all their inventions to their employer. In such cases, if all the inventors have a common employer, the decision about filing may be taken by the employer. However, where one or more of the inventors is not under any obligation to assign the invention, it is desirable to obtain permission before filing. In the UK this is not strictly necessary, since a UK patent application can be made by anyone. The application can however only be granted to persons legally entitled to it: therefore between application and grant, arrangements must be made to transfer rights in the application to those entitled to the invention. In the USA, an application can only be made in the name of the inventors: if the application is to belong to the employer, it is assigned after filing.

Security

Strangely, security considerations must be borne in mind for all patent applications first filed by UK nationals or US residents. In order to safeguard information of potential importance to national security, both countries require that no patent application may be filed overseas without special permission until a specified time (six months for USA, six weeks for UK) after a patent application has been filed locally. Such permission is readily obtainable, where desired or necessary (for example, where there are inventors on both sides of the Atlantic, permission will be required from one country before a first filing in the other) but should be obtained in advance. Permission is much more difficult to obtain retrospectively (in UK, usually impossible). Penalties for breach of this regulation are fines or imprisonment (in the UK - not usually inflicted unless the application really does disclose defence-sensitive information) or invalidity of the subsequent patent (in the USA).

Professional Services

It is highly advisable to have the application drafted by a qualified and experienced attorney, to ensure it meets appropriate standards of disclosure and provides a sound basis for proper broad protection. This may take a little time and significant expense - but this is an area where it is foolish to economise.

9.6.2 Overseas filing

The initial application gives a year's priority in most countries of the world. Any subsequent application in any of these ("Convention") countries made within one year of the original filing will be entitled to the priority of that original filing, for matter disclosed in it. Any intervening filing or publication of such subject-matter will be considered to be pre-dated by the original application. This is a very valuable privilege for the patent applicant. It gives a further year for further evaluation of the technical and commercial prospects of the invention, before the considerable expense of overseas filing has to be undertaken. For applicants who do not have the resources to develop an invention for themselves, it gives some time to obtain backers or licensees. Further, additional developments of the original invention can now be added to the overseas filings: though priority is only allowed for what was originally disclosed.

Formerly, a local national filing had to be made in each country where it was desired to protect the invention. Failure to file in any country made it likely that all opportunity to obtain protection would be lost there. New systems (EPO, PCT - see below) have considerably softened the harsh choices that formerly had to be made about eleven month after filing the priority patent application: but it is still wise to think carefully at this stage to plan where protection for the invention is needed. Patent filing costs do not vary much between countries (perhaps by a factor of 10 at most between dearest and cheapest). The value of any resulting patent can vary much more: according to the size and economic wealth of the country where the patent is filed, according to the type of patent protection available for the subject-matter of the invention, and according to the efficiency of the legal means available for asserting the patent against infringers. A long filing list rapidly demonstrates the law of diminishing returns.

Great care must be taken in foreign filing to ensure the papers reach the overseas country in time to be filed at the local Patent Office on or before the anniversary of first filing. Many official terms for reply set by Patent Offices are extensible -- but not the term in which priority must be claimed under the International Convention. Failure to file within the year means loss of priority: often this means loss of rights, where the invention has been published within the priority year, either by the inventor or a rival. The only case in which extension is normally allowed is if the Patent Office is shut (on a weekend or national holiday) on the date on which the period expires: then the period is extended to the first following date on which the Patent Office is open.

Publication

In most countries (Europe; Australia; Canada; Japan, for example - but not USA) the patent specification is made public eighteen months after the original priority date. At this stage the public can buy copies, and learn what the patent applicant claims as new, and how to put it into practice. At this stage, the patent applicant has no enforceable rights to stop the public using the invention in any way. However, if the patent is subsequently granted, the patentee can then claim "reasonable compensation" in respect of any intermediate use covered both by the claims as originally published and as granted. In USA it is considered unreasonable to publish the inventor's idea until it is known what rights (if any) will be given in return: thus the USA is the only country which still offers a genuine choice between keeping secret and patenting. Other countries offer only the choice between keeping secrets and the opportunity to attempt to patent. Furthermore, unless the US inventor is prepared to run the risk of losing foreign rights, the invention will be published overseas in any case at 18 months from priority.

Patent prosecution

This normally follows a request from the applicant after publication. A search may be carried out by the patent office to help the applicant decide whether it is worthwhile to file such a request. The period for request varies from 6 months from publication in the EPO to seven years from filing in Japan. In USA, examination is automatic for all patent applications that are filed. Patent prosecution consists in arguing with the Patent Office Examiner about the form and substance of the patent to be issued, if any. Most discussions will be about whether the invention (as defined by the claims) is new and not obvious: but the Examiner may also object to the description and claims as incomplete or ambiguous, and require clarification. In reply, the applicant will argue the points made by the Examiner, or accept them in whole or in part, and will often propose amendments, particularly to reduce the scope of the claims to what is more clearly new and inventive. A particularly difficult objection is that the description of how to carry out the invention is incomplete and does not enable the person skilled in the art to repeat it. This has to be argued against: it may not be dealt with by amendment, because this would mean "adding subject-matter". The rule is that the specification must contain an adequate description when it is filed, and this cannot thereafter be supplemented.

role of Patent Office Examiner

Grant

Prosecution normally terminates either in allowance of some claims or refusal of the application as a whole. In the latter case, it is normally possible to appeal to an Appeals Board within the Patent Office (sometimes to an external tribunal). If claims have been allowed, the patent is granted (in USA, publication of the patent specification now takes place for the first time). Now opposition by third parties may be possible. In the European Patent Office, for example third parties have a period of nine months in which to register objections to the granted patent. These objections are heard before the Opposition Board of the EPO, and both patentee and opponent have full opportunity to put their case. The Board then decides: either to grant the patent unamended, to revoke it or (frequently) to grant it with claims amended to take account of, at least , some of the opponent's objections. Appeal is then possible to the EPO Board of Appeals, whose decision is final. In Japan, opposition takes place before grant: this practice is however unpopular with patentees as opposition proceedings (like all litigation) are very slow, and the patentee has no way of enforcing any rights until the patent is granted.

appeals procedures

Renewal fees

After grant (indeed in some cases before) renewal fees are payable to keep patents in force. Most frequently these are annual (though every three and a half years in USA) and they tend to increase steeply as the life of the patent goes on. This is reasonable - if the patent is useful, it will be earning good profits towards the end of its life as the market becomes fully developed - if not useful, the patentee should be discouraged from maintaining it. Finally the patent comes to the end of its normal term and expires. However, some countries make provision for extension of patents for certain reasons, including war loss, and great and unrewarded merit. More recently, inventions where

expiring and extension of patents

stringent safety regulations prevent early marketing (for example pharmaceuticals and agrochemicals) have been the subject of special legislation allowing partial extensions. It is possible that similar arrangements may be made for biotechnology inventions in the future.

9.7 Stopping infringement

How does the patentee enforce the monopoly the patent gives? A patent is a private right. In most countries patent infringement is a civil rather than a criminal matter, comparable to breach of contract rather than theft. The patentee must enforce his own right - the government will not do it for him.

patent enforcement Patent enforcement is (ultimately) by Court action. The patentee must sue the alleged infringer in the Courts. However, taking such action should be a last resort, rather than an automatic option, for two main reasons. These reasons apply to all litigation, in some degree, but more especially to patent litigation.

9.7.1 Uncertainty

The result of the litigation is particularly difficult to predict. The defendant will argue (almost invariably) that his product does not infringe the patent and that in any case the patent is invalid. If either argument is successful the patentee will lose his case. There is nearly always a case to be made that a patent is obvious, and (as discussed above) the strength of such arguments is particularly difficult to assess. If the patent is held invalid by the Court, the patentee loses all his rights - including the right to receive royalties from any existing licensees. Thus the patentee often has considerably more at stake than the defendant, and this must be taken into account in deciding whether to sue.

9.7.2 Expense

patent litigation Most litigation is expensive. Patent litigation tends to be particularly so, because of the complex legal and technical issues involved. This is particularly so in common law countries, such as UK and USA. A contested patent infringement action in USA is unlikely to cost less than $2M. These expenses will rarely be recovered in full even if the patentee is successful. In UK, a losing defendant must make a contribution to the plaintiff's costs, but this will be based on the theoretical minimum cost of bringing the action. In USA, unless the plaintiff can prove that infringement was wilful and deliberate, no legal costs can be recovered.

9.7.3 Procedure

This varies from country to country. Generally, infringement and validity are considered together, and part of the defendant's case is a counterclaim for revocation of the patent on the ground of invalidity. In common law countries each party has to produce the documents it holds that are relevant to the issues in dispute. In USA, particularly, this procedure (known as 'discovery') is so extensive that it is a prime cause of the expense of patent actions. The documents revealed in this procedure are often crucially important in determining the outcome of the case. If the defendant is alleging that the invention is obvious, his case can be totally undermined by memos written by his scientists saying for example, that the newly published patent solves a problem that they had long been wrestling with. Or instructions by inventors to conceal their preferred way of carrying out the invention may be fatal to a US plaintiff.

9.7.4 'Saisi'

In the civil law countries of continental Europe, discovery of documents is not provided for. France and Belgium, however, provide another means to enable the patentee to establish infringement. The patentee may employ a public bailiff to attend the site of a suspected infringement (such as a factory or warehouse), to enter it and take copies of documents and samples of infringing products. On the basis of the results of this procedure - termed a 'saisi' - the plaintiff then has a very short term (for example 15 days) to decide whether to continue with the action.

9.7.5 Remedies

Where a successful plaintiff establishes his patent is valid and infringed, the Court may order the defendant to:

- stop infringing;

- pay damages for past use (or royalties);

- account for profits;

- destroy infringing articles.

Generally, the only remedies that the plaintiff is entitled to, as of right, are damages. Other remedies are discretionary, and will only be ordered if the Court thinks fit. For example, delivery up, or destruction, of infringing goods may not be ordered where they can readily be modified to be non-infringing. Damages are determined by asking the same question as is asked in the case of other invasions of the plantiff's rights: what loss has the plaintiff suffered as a consequence of the defendant's wrongdoing? This will vary according to circumstances. If the plaintiff licenses his invention freely to all *actual remedies depend on circumstances* comers at a standard royalty, the damages are simple to calculate: they are simply the royalty that the defendant would have paid had he been licensed. More commonly, damages will be higher than this. For example, the defendant's infringement may have raised doubts as to whether the patent can be asserted. In consequence, the plaintiff may have had to accept a lower royalty from all his licensees. If this can be shown to be due to the defendant's conduct, the defendant might have to make up the difference between what the plaintiff has actually received and what he would have received. A more common case is where the plaintiff has not granted licences, but exploits his invention by selling exclusively. In this case, the defendant may be treated as having deprived the plaintiff of sales: and will then have to pay the plaintiff the profit the plaintiff would have made on those sales. If the defendant's competition has made the plaintiff reduce prices, the defendant may have to pay the plaintiff's notional profits on the higher price (though there may be some reduction if it can be shown that the lower price has expanded the market). Sometimes it may be difficult to assess what the plaintiff's damages are, but once infringement has been proved, damages tend to be assessed with a broad sympathy to the plaintiff - the defendant, after all, has been found to be in the wrong. If the plaintiff prefers, he can ask the defendant to account for (and pay over) the profits made from his infringing act - but this remedy is an alternative to damages, not in addition to them.

9.7.6 Injunction

The most powerful weapon in the hands of the Court is generally the injunction to stop further infringement. This remedy is discretionary, but will normally be granted to a successful plaintiff unless the defendant shows special circumstances (for example the patent is on the point of expiry, or the defendant has sold his business and is no longer in a position to infringe). A defendant who is subject to an injunction is bound by the Court not to infringe. If he does, he may be fined or sent to prison, as well as having to pay further damages. In many cases, the injunction is the main remedy that the plaintiff seeks - for example, if the infringement is just beginning or even has not yet started (an action against a threatened rather than an actual infringement is termed a *quia timet* action).

preliminary
(interlocutory)
injunction

A final possibility for the plaintiff is to ask for a preliminary (also called an interlocutory) injunction. Patent actions often take years. If during this time the defendant can carry on selling, the damage caused to the plaintiff may not subsequently be fully compensated by damages (either because the market develops in a way unfavourable to the plaintiff, or because the defendant is simply unable to pay in full). In such a case, the plaintiff may ask the court for a preliminary injunction. The plaintiff must show there is an arguable case of infringement, but is not required to prove it in full. The Court will then consider the balance of convenience - is the damage that the plaintiff will suffer if the infringement is allowed to continue until the court's final decision greater than that the defendant will suffer if he has to stop now and subsequently proves that what he was doing was entirely legitimate? Of course, the plaintiff has to undertake to compensate the defendant for his losses if infringement is not found. It might be thought that such preliminary injunctions would rarely be granted, but this is not so. Further, they often result in the termination of the action, particularly where a powerful plaintiff sues a small defendant. A defendant who may be unable to sell his product for a further two years or more may have little incentive to continue expensive and uncertain litigation, and will settle on the best terms he can get.

9.7.7 Appeal

A final decision of the Court will normally be open to appeal. The Appellate Court concentrates on considering points of law, rather than fact: it will not substitute its own decision of fact for that of the trial judge, unless it can be shown that there was no reasonable basis for his decision. A further appeal may be possible, with leave (in the UK to the House of Lords, in the USA, quite exceptionally, to the Supreme Court). Such appeals can add substantially to the time and expense of patent litigation, as well as the difficulty of predicting the results.

9.8 International Systems

The phrase "World Patent" is sometimes seen on gadgets. There is no such thing. Patents are, in principle, national rights: and to obtain a patent in any country it is generally necessary to apply for one in that country. Nevertheless there are international systems which modify this principle in some degree: and increasingly there is a trend to international harmonisation. This is no recent phenomenon: the list of international conventions starts in 1884.

9.8.1 The Paris Convention

International
Convention of
Patents

The earliest of the patent conventions is sometimes known simply as the International Convention on Patents but more often as the Paris Convention, after the city in which the first treaty was signed. Since 1884 there have been a number of revisions (again known after the cities in which the negotiators met - the Hague, London, Stockholm). But the main features of the convention were clear from the beginning. These are the priority right and the principle of national treatment.

Priority Right

This is perhaps the most fundamental feature of international patent law. Under the Paris Convention, each signatory nation agrees to recognise a first patent application in another member country as giving rise to a right of priority. The period of the right is one year. During that year, the applicant (or his assignee) may apply in any other member country for a patent on the same invention and have his priority right recognised. That is to say, what has been disclosed in the earlier application is treated as having the date of the earlier filing: disclosure after the earlier filing, or even a subsequent application in the second country, cannot be used to invalidate the later application.

Convention
Year

This is an extremely valuable right. A single patent application in a single country is not cheap: protecting an invention in major countries of the world is correspondingly more expensive. The "Convention Year" (as it is often known) provides the inventor the opportunity to perfect the invention, to test the market and to seek backers. A year is little enough for all this.

Note that the right is only given to the first filing (by any particular applicant). It is not in order to file applications on the same invention at three-monthly intervals, waiting until someone else publishes the invention, and then claiming the priority of the latest application with an earlier priority date. A priority application may however be abandoned - either formally, by filing an irrevocable declaration to that effect, or by allowing the convention year to expire. If no rights are claimed or outstanding, the application can then be refiled, and the priority year starts again.

National treatment

The other fundamental principle of this Convention is the principle of national treatment. It is easy to see that it is in a country's interest to encourage its inventors to disclose their ideas in return for a limited monopoly. It is less clear that it is in the country's interest to grant similar rights to foreign inventors, who may publish their ideas overseas and hence make them accessible there without any further encouragement: and may perhaps exercise their monopoly only by using it to protect imports. If every country takes this view, there will be a much reduced incentive for inventions to be published and exploited. Nevertheless there is a great temptation for countries to treat foreigners less favourably that their own nationals. In the latter half of the nineteenth century, such discrimination was seen as only reasonable. It was therefore particularly foresighted of the negotiators in Paris to lay down the principle that each country would treat the nationals of other countries in the same way as they treated their own. This has been a very useful foundation for further international co-operation.

Other provisions

These include in particular norms on working requirements. These have often been severe in developing countries, which may have seen them as a way of limiting the unpleasant necessity of granting rights to foreigners. The current text of the Convention requires a period of at least three years from grant to elapse before any sanction can be imposed for non-working: and the sanction that can then be imposed is limited to the grant of a compulsory licence. Revocation cannot be granted for a further two years, and then only if the position cannot be remedied by the grant of compulsory licences.

Membership

Most of the world's countries are members of the International Patents Convention. Major exceptions are India and Pakistan, and South American countries other than Argentina and Brazil. Even Argentina and Brazil are only members of earlier versions of the Convention.

9.8.2 European Patent Convention (EPC)

Strasbourg
Convention

In 1963, a number of European countries signed the Strasbourg Convention, which was intended to bring European patent laws more closely in line. This was followed in 1973 by the much more significant European Patent Convention.

Objectives

European
Patent Office

The EPC builds on the Strasbourg Convention by setting a framework of law for member countries. For example, it lays down what is patentable and what is not (see below) and sets a uniform patent term of 20 years. Most importantly, it established a central European Patent Office (based in Munich) whose function is to receive, examine and grant European patent applications. Such applications are not, however, for a unitary European patent: they are for a bundle of national patents. Each application to the European Patent Office must designate the European countries in which protection is required.

Procedure in the European Patent Office (EPO)

fees

An application with a specification (and drawings if appropriate) is filed with the Munich Office. The specification may be in any of the official languages of the European Patent Office, (English, French or German). The applicant must select the member countries in which protection is required and pay a small designation fee for each. Priority of an earlier filing may be claimed under the Paris Convention. A search is carried out (usually by the search branch of the Office, based in The Hague), and the results communicated to the applicant, typically before the application is published. Publication takes place at 18 months from the earliest priority date claimed. The applicant then has six months in which to request examination. If the invention is still of interest, and the search has not produced anything too damaging in the way of prior art, the applicant then requests examination and pays the examination fee. There is then a pause of maybe a year or more, until the application is taken up for examination. An Examiner (part of a three-person examining team) reviews the application, and sends a letter to the applicant detailing objections. The applicant is given a four-month term to respond (which can be extended for good reason) by amending the specification to meet the objections or arguing that they are not well based. Technical evidence may be filed if required. Sometimes prosecution is assisted by arranging an interview with the examiner. The applicant's reply may require a further reply from the Examiner, accepting or rejecting (in whole or in part) the points made. If acceptable, the text will be finally agreed (this is a last opportunity to make minor corrections) and then reprinted for grant. At this stage, the applicant has to provide translations of the allowed claims (not the specification) in all three official languages of the EPO.

If the applicant and the Examiner cannot agree, the Examiner will reject the application. The applicant then has the opportunity to take his case to the Appeal Board of the EPO. A Board of three senior experienced officials will be appointed to review the case and to hear the applicant's arguments. If they uphold the original Examiner's view, the case is finally refused. There is then no possibility of further appeal unless the case is a quite exceptional one (perhaps, for example, if it could be possible to make a case to the European Court that the EPO had exceeded its powers).

objections to the granting of European patents

After grant of the European patent, competitors have a period of nine months in which to file objections, if they wish. Such objections are made by filing an opposition at the European patent office, setting out the grounds on which it is alleged the patent should not have been granted, and paying the official fee. The applicant replies to these objections, and the opponent may reply further. In due course, a hearing is appointed before an Opposition Board, at which both patentee and opponent are entitled to appear and argue their case. The Opposition Board gives a decision: which may then be appealed by the losing side to the EPO Board of Appeals whose decision is final. If the patent is revoked, it no longer has effect in any of the countries of Europe for which it was designated.

After grant, each European patent has effect as a national patent in each country designated. The patent must be kept in force locally by payment of local renewal fees. It may be challenged in the local courts, and perhaps revoked: but this will not affect its status in any other country. There is one expensive formality to be fulfilled before the patent is effective in the designated country: except in Germany, the specification must be translated into the local language. If this is not done within the specified term, the patent lapses irretrievably for the country in question.

Members of the EPC

A full list of EPC members is given in Table 9.I. Membership is not limited to EC countries (and indeed until recently did not include such countries as Greece, Portugal or Ireland). Non- EC members of the EPC are Austria, Sweden and Switzerland. East European Countries (Poland, Czechoslovakia, Hungary) may well join the convention in the near future.

Table 9.1
The Patent Cooperation Treaty Member States
1 January 1993

Australia	Austria*
Barbados	Belgium*
Benin	Brazil
Bulgaria	Burkina Faso
Cameroon	Canada
Central African Republic	Chad
Congo	Czechoslovakia
Denmark*	Finland
France*	Gabon
Germany*	Greece*
Guinea	Hungary
Italy*	Ivory Coast
Japan	Korea (North)
Korea (South)	Liechtenstein*
Luxembourg*	Madagascar
Malawi	Mali
Mauritania	Monaco
Mongolia	Netherlands*
Norway	Poland
Romania	Russia
Senegal	Spain*
Sri Lanka	Sudan
Sweden*	Switzerland*
Togo	Ukraine
United Kingdom*	USA

* Members of the European Patent Convention

Patentable subject matter under the EPC

Member countries have to recognise common standards of what can be patented. It is compulsory to allow patents to chemical compounds and pharmaceuticals (in the '60's, this was a radical change for several countries). However, countries who join the convention are generally allowed a period in which to adjust their law. DNA is a chemical compound - so claims to genes are allowable in principle as discussed above, (always provided they meet other criteria, such as novelty and non-obviousness). On living material, the convention provides:

- plant and animal varieties are unpatentable;

- essentially biological processes are unpatentable this does not apply to microbiological processes or the products thereof.

rules of the
Board of
Appeal of EPO

What does this mean? So far, national courts have not considered these provisions in detail. The Board of Appeal of the European Patent Office has, however, been active. Its cardinal rule has been that exceptions to patentability are to be construed narrowly, so as to make as much patentable as possible. Exceptions to such exceptions are correspondingly construed widely. Thus, a plant variety has been held to mean that which can be protected under a plant variety right (something quite specific, see below) and hence not to include a novel plant genus (or family of varieties). There is some doubt about what an animal variety is since there is no scheme for protecting varieties of animals corresponding to the UPOV convention. (UPOV = Unions for the Protection of New Varieties of Plants). However, the EPO has allowed a claim to a new class of animals: in a decision rather misleadingly referred to as the Harvard Mouse case (the animal exemplified was a mouse, but the claims were to non-human mammals - even claim 11 covered rodents broadly).

'Essentially biological' rules out customary methods of selective plant and animal breeding (in any case difficult to patent because not new and not repeatable) but does not exclude such processes which have been modified by some technical intervention. Thus a process for breeding hybrids in which one of the parent lines was maintained by cell cultures was held not to be essentially biological, and hence to be patentable (though the process subsequently turned out not to be new).

9.8.3 Community Patent Convention (CPC)

This Convention originated at the same period as the EPC, but has made much slower progress. It has yet to come into force. It is confined to members of the European Community. When in force it will:

- provide a unitary patent for the whole community;

- be maintained by one set of renewal fees;

- promote the growth of a Community-wide patent regime, by means of a common appeal Court.

difficulties with
ratifying CPC

Difficulties with the convention have been: constitutional problems in ratifying it (Ireland, Denmark) and the expense. This arises from the requirement for the text of the specification to be translated into all languages of the Community. The sanction against patentees who do not do this (within a period of three months from grant) is invalidity throughout the Community. Similarly, the renewal fees payable will be less than the sum of fees payable for independent patents in all EC countries - but without the option to select protection in some of the countries only and save money. There are grave doubts whether even the richest patentees will be willing to undertake such costs on a routine basis. No clear date for the CPC to come into force has yet been set - and there are few demands for it to do so, as it offers few advantages over the very successful EPC.

9.8.4 Patent Cooperation Treaty (PCT)

The phrase "World patent applied for" is a well-known misnomer - but since the advent of the PCT it can be more nearly true than formerly. The Patent Cooperation Treaty, introduced in 1978, is a world-wide system for unifying patent application, patent novelty search, and the first stage of patent examination. However, unlike the EPC, it does not actually grant patents: this task is passed to member countries.

Membership

Fifty-two countries are currently members of the Treaty - the full list is given in Table 9.1. All major developed countries are members, and many less developed countries. Regional Patent Offices, such as the EPO, are also members, and protection in such offices may be sought through the PCT.

Procedure

A PCT application may be filed at any of a number of nominated receiving offices (UK, EPO, USA, Japan among others). The application is made in the language appropriate to the office. Priority may be claimed under the Paris convention. The applicant must designate the territories in which protection is required and pay appropriate fees (including a fee for each country designated up to a maximum of ten - so all countries can be designated for the same cost as ten). The claims are searched for prior art (this has to be done within three months) and the results communicated to the applicant. This allows time for the application to be withdrawn (should the applicant so decide) before publication takes place at 18 months from the earliest priority date claimed.

Chapter II

The applicant then has a choice, for most countries. The majority of countries have ratified Chapter II of the Treaty, which provides for international examination on the basis of the search. For those countries, the applicant may elect international examination, by filing the appropriate request and paying the fee. This only applies to countries which the applicant originally designated. Election must be made before the expiry of 19 months from the priority date. The examining patent office will then issue a report on patentability of the invention, assessing novelty, obviousness and any formal defects. The applicant may reply to the report (if negative) and amend the claims or argue the relevance of any of the citations said to affect inventiveness. The examiner will then issue a final report, maintaining the original opinion, modifying it or withdrawing it.

The national phase

The national phase of the procedure must be entered by 20 months from priority (in those countries that the applicant has not elected international examination) or by 30 months from priority (in countries for which such examination has been elected). To enter the national phase, the applicant must file an application in the selected country, in the local usual practice of the country. Countries do not regard the result of international examination as binding - sometimes indeed they take little notice of it.

Advantages of PCT

The primary advantage of the PCT procedure is that it enables the postponement of the expense of overseas filings for up to a further 18 months beyond the 12 months that the Paris Convention allows. That is a total of 30 months from the original priority date. During this period the applicant receives a full search of the invention and an assessment of patentability. There is further time to assess the merit of the invention and perhaps find a licensee. The major expense of a wide foreign filing is postponed until much more information is available to judge the merit of such an investment.

Disadvantages

The main disadvantage is the extra cost - but this, equivalent to about one extra national application, is not much in a large filing, and fully compensated for by the extra flexibility. There used to be a number of traps in the PCT procedure, but few of these remain. One point to be remembered is that if protection is required in Poland, a translation must be filed at 20 months, whether or not international examination is requested.

OAPI

Another regional patent office, for French-speaking Africa, is OAPI based in Cameroon. Eleven African Countries are members - the full list is Benin, Burkina, Faso, Cameroon, Central African Republic, Chad, Congo, Gabon, Mali, Mauritania, Senegal and Togo. These countries may be protected by a single patent application - conveniently filed through PCT.

9.9 USA vs Europe

It is interesting to compare and contrast the patent law of the USA with that of Europe. US patent practice differs from the rest of the world in a number of respects. These may be summarised as follows:

	Europe	USA
Who gets the patent?	first to file	first to invert ("interference")
Novelty	inventor must not publish before filing	.... must not publish more than a year before filing
When is specification published	at 18 months	no publication before grant
Speed of grant	slow (c. 5 years)	very variable (1 to 10 years or more)
Term	20 years from filing	17 years from grant
What can be patented?	some things unpatenable (eg plant varieties)	almost everything patentable (except human beings)

Note: In most respects, the rest of the world's patent systems are more like those of Europe than that the USA. Some changes may be in store (see Patent Harmonisation, in Section 9.15).

A note on Interference Practice

Surely it is fairer that patents should be granted to the actual first inventor, rather than merely to the first person to file a patent application. In theory, perhaps - but the practice is more difficult. The act of invention is difficult to define (see the discussion on obviousness, above). This makes dating it very difficult. The American system, though fairer in theory, in practice is artificial and legalistic. How does it work?

Interference

When the US Patent Office recognises that two applications (or an application and a patent) are claiming essentially the same invention it will declare an interference. This is a procedure for determing who is the first inventor, and hence entitled to the patent. What criteria are used?

Criteria for Inventorship

There are three: Conception, Reduction to Practice, and Diligence.

Conception

Conception is the complete idea of the invention in a workable form. The objective of the invention is not sufficient, unless the means for carrying out the invention is also apparent to the inventor.

Reduction to practice

Reduction to practice is the physical act of carrying out the invention. This may be on a small scale, and quite crudely and inefficiently, but in a way sufficient to demonstrate that the invention works as the inventor proposes. Filing a patent application with a proper description of the invention also counts as reduction to practice - this is "constructive reduction to practice'.

Diligence

Diligence is what the inventor does in seeking to convert his invention from an idea into a working form - to proceed from conception to reduction to practice. For example, if the invention is a machine, it may include having drawings made, and parts ordered - even phone calls to tardy suppliers.

Two other points should be noticed:

- all acts of inventors must be independently corroborated;
- only acts done in the USA count.

rules relating to diligence   Both these rules arise so that the tribunal can have reasonable assurance about the quality of the evidence before it. It is obviously a temptation to an inventor, who, perhaps, has not kept very good records, to try to improve them so that they show what (he is confident) really happened. Equally, while the second rule may seem a slur on other nations, it seems to have been brought in because of doubts that foreigners would respect laws against perjury that would be difficult to enforce against them in practice.

An interference may be between two or more parties. The first to file a patent application is the senior party: the others are junior parties and bear the burden of proof to show that they made the invention before the senior party.

Who wins?

The winner is the first to conceive and the first to reduce to practice.

If the first to conceive is the last to reduce to practice, then he wins only if he can prove that, from a date before the other's conception, he was continuously diligent in seeking to reduce the invention to practice.

Put another way, the first to reduce to practice wins (remember that this may include filing a patent application) unless the other party shows continuous diligence in reducing his invention to practice from before the first party's conception.

Is this a fair rule? Clearly it is not the only possible rule. Other rules might be simply to give the invention to the first to conceive (Canada used to do this, but now has adopted the first-to-file rule) or the first to reduce to practice. But perhaps this would have little advantage over giving the patent to the first to file. The rule recognises that patents are a reward for introducing new technology, and tries to balance the contribution of the idea and of its practical introduction.

There is one powerful objection to the rule. It is not transitive (in the mathematical sense). In sporting competitions (football or chess) it is fairly common for team A to beat team B, team B to beat team C and team C to beat team A: but we do not expect this to happen in a contest for legal rights. Yet it is possible (at least in theory) in an interference. Consider the following situation:

A conceives and reduces to practice in June. B conceives in February and reduces to practice in August. B is not diligent. C conceives in March and from that date is continuously diligent until reduction to practice in September.

Here A beats B (earlier reduction to practice): B beats C (for the same reason). But C beats A (later reduction to practice, but earlier conception combined with diligence. Who made the invention first?

There seems to be no record of this situation actually occurring in practice. What is said is that, if it did, and there was a proceeding between the three parties, priority would be awarded to the party who beat the senior party. In the above example, if A is the first to file a patent application, and hence the senior party, C will win. This is on the theory that a three-party interference is really only two two-party interference, and the two junior parties are not in competition. However, this answer, based on procedural law, hardly provides the conviction that the fundamental legal basis of the rule is a just one.

There has been talk for many years of the USA changing from a first-to-invent to a first-to-file system. If current discussions on global Patent Harmonisation come to fruition, this may well happen soon. Even if so, interferences will likely continue for some years between patent applications filed before the reform takes place.

9.10 Plant variety rights (PVR, PVP)

formation of UPOVThe Plant Variety Right is a special right available to the breeder of a new plant variety. The plant breeder is in the typical position of an intellectual creator, that his work is expensive to create, but easy to multiply. The right grew out of the conviction (now sometimes challenged) that the patent system is unsuitable for protecting plant varieties, for two reasons: it is not always clear that such varieties are inventive (non-obvious): and it is not generally possible to describe how to repeat their creation. Accordingly, a number of countries got together to form the Union for the Protection of

New Varieties of Plants (UPOV). This body, based in Geneva and affiliated to the World Intellectual Property Organisation, currently has twenty-one countries as members (for a full list see below). Its conventions, regularly updated (most recently in 1991) set the principles for the protection of plant varieties in member countries.

UPOV Members - June 1992	
Australia	Belgium
Canada	Czechoslovakia
Denmark	France
Germany	Hungary
Ireland	Israel
Italy	Japan
Netherlands	New Zealand
Poland	South Africa
Spain	Sweden
Switzerland	United Kingdom
USA	

Several other countries are expected to join soon.

The following discussion relates to the law as it is in effect at the date of writing. The 1991 convention has proposed a number of significant changes, but these have yet to be introduced to the legislation of any member country.

9.10.1 What is a plant variety?

A plant variety (sometimes also referred to as a cultivar) is a genetically uniform subspecies of plant, fulfilling three main conditions: it must be DUS.

- distinct;

- uniform;

- and stable ('DUS').

Distinctness

To be distinct, the variety must differ "in at least one important characteristic" from all known, or previously registered, varieties. Note that this characteristic need not be unique - and rarely is. A new variety of wheat V may be made by crossing a known variety P with a short stalk and a known variety Q with a narrow leaf, and selecting from the progeny a new variety with both characteristics. This then is distinct from the first parent by having a narrow leaf and from the second by having a short stalk. Note also that the importance of the characteristic need not be economic. In the example, a short stalk is very important economically, because the wheat is less prone to fall over ("lodge") in bad weather: while the narrow leaf may have no significance to the usefulness of the plant. But both equally can make the variety distinct.

Uniformity

A variety is a group or race of plants. For the variety to be uniform each member of the group must be substantially the same. The requirement is not absolute, and indeed slightly different criteria are applied in different species: but broad uniformity is necessary. Environmental conditions may of course affect how each seed grows, but if the field is uniform, each seed should come up looking the same as its neighbour. If, instead of crossing varieties P and Q, we mix their seed for sale in the same bag, we have not made a new variety: and the new blend does not qualify for plant variety rights.

Stability

Not only must the plants look the same throughout the field: they must look the same when replanted the following year. This is the requirement for stability. It is not, however, necessary that the seeds of the variety breed true. Stability may be obtained in other ways - for example F1 hybrids of consistent properties are obtained by crossing parent lines which themselves breed true. Maize hybrids are a common example. Other plants may be reproduced by vegetative propagation (eg potatoes).

To qualify for protection, a variety must also be novel - that is to say, it must not itself have been commercially exploited prior to application for registration. Novelty requirements differ from country to country, particularly in the period allowed for prior exploitation overseas.

9.10.2 Procedure for application for a plant variety

As with patents, application must be made to each country where protection is required. Official examination varies from country to country. In USA the applicant files a detailed description of his variety, and a paper comparison is carried out of this with known varieties. In Europe, the variety is grown in official trials (in UK for example, barley will be grown at three different locations over two years). For agricultural crops in Europe, this examination forms part of official trials to determine if the variety should be placed on official lists of varieties that may be legally sold. In Australia, the applicant must submit results of trials carried out by him or on his behalf. As part of the examination procedure, the applicant must propose a name ("Golden Promise" say, for barley, or "Elizabeth of Glamis" for a rose) by which the variety will be known. This is a common name rather than a trademark: but it is illegal to use the name except for the variety in question, which affords a valuable means for the breeder to assert his rights.

9.10.3 Rights of the breeder

What rights does the breeder receive? Broadly, the right is more limited than that of a patentee. The breeder gets the right to exclude others from multiplying the variety for selling as reproductive material. But the grower of the seed can sell it freely for consumption.

breeder's rights   There are two significant exceptions to the right. The breeder's exemption is the right to use a protected variety as a starting point to develop a new variety. If the resulting new variety is distinct, the owner of the protected variety has no rights against it. But it is not possible to use a protected variety as part of the normal production cycle of a new variety - for example, as one parent of an F1 hybrid.

farmer's privilege
The farmer's privilege is the right to multiply seed for his own further use. The exact scope of this right is not codified in the convention, and differs from country to country (in USA, for example, a farmer may not only save for his own use, but may sell for replanting up to half the seed he produces). This exemption is of great concern to breeders: in Europe, use of farm-saved cereal seed varies from 20% up to 80% in some countries.

PVP is highly specific
The protection given by PVP (Plant Variety Protection) is highly specific - one right cannot infringe another, and only a small change is necessary to get a separate registration. Nevertheless it is useful and relatively easy to enforce (especially in highly regulated European agriculture). One perceived advantage is that lawyers are less involved than with patents, with less opportunity for trouble and expense.

9.11 Patents vs Plant Variety Protection rights

It is sometimes maintained that patent rights for plants are unnecessary, because plant variety rights offer all the protection that is necessary or desirable. The converse is also suggested on occasion. Is either view right?

The answer is that the patent system and the plant variety right scheme have different scope and functions. **Both systems are needed**.

Patents are for genetic plant technology. Usually this technology is at a very early stage of laboratory development, requiring several years' further work in the laboratory and the field before any product is ready for the market.

PVP rights are for specific varieties ready to be sold. PVP rewards the breeder's useful and laborious work, so easily copied, but often not "inventive". In the example given above, it is clearly obvious that varieties P and Q can be crossed, and that some of their progeny will have both narrow leaves and short stalks. This should rule out granting a patent for the variety V - only PVP rights are obtainable.

Equally, genetic technology is incapable of protection by plant variety rights, and can only be protected by a patent. In fact, many biotechnology developments coming to the market will be protected both by patent and by PVP. The original genetic invention (use of a specific bacterial gene to give insect resistance, say) will be patented. This gene will then be inserted into a range of new varieties. Because the breeding process is slow and time-consuming, this process will take several years. Each of the new varieties will be protected by PVP when it is introduced to the market.

9.12 Trade secrets

Are there any other ways in which the biotechnologist can protect the results of his research? Someone who makes a discovery is not bound to reveal it to the world - instead, he may keep it to himself. As long as no-one else knows of it, he has the exclusive benefits of it.

What are the advantages and disadvantages of doing this?

1)   The protection depends on secrecy being maintained. The invention cannot be copied, not because some right is infringed, but because competitors do not know how. If selling the product gives away the secret, no protection is obtained: likewise if the product can easily be multiplied (as with bacteria or true-breeding seeds). Also, protection may be lost as a result of the secret being discovered independently by someone else.

2)   Protection lasts indefinitely. So if the secret is sufficiently difficult to work out, protection may continue as long as the relevant technology is in use.

3)   The secret may be licensed to others, if they are prepared to bind themselves to maintain the necessary secrecy.

The holder of a trade secret needs to take care to preserve it in confidence. It should be disclosed only to employees who need to know it: and care should be taken that they are aware of the status of this knowledge and of their obligation to keep it secret. An employee who leaves his job may not pass the secret on to his next employer. If this obligation is ignored, it may be possible to prevent the new employer from using the secret information. However, where the information was obtained by a third party in good faith (that is to say, not knowing or having any reason to suspect that the party disclosing it was not entitled to), its use cannot be stopped - all the owner can do is sue the party who wrongfully disclosed it.

Trade secrets are, in principle, an alternative to patents - you cannot have both for the same subject-matter. To be granted a patent, it is essential to disclose a method (in USA the best method) for carrying out the invention. Of course, it is possible to patent the original idea and keep secret subsequently developed refinements. They can be combined with plant variety rights in some cases - specifically in the case of hybrids. Parent lines for maize hybrids are nearly always maintained as proprietary trade secrets. However, increasingly they are also being protected by plant variety rights - this is possible since there is no obligation to make the subject of a PVP available to the public.

A very important use of trade secrecy is to protect commercially important strains of micro-organisms. Many improved strains (used, for example, to make antibiotics) are kept confidential in preference to patenting. This is in part because (except in USA) an unsuccessful attempt to patent results in the strain becoming freely available.

9.13 Other Possibilities

Few other intellectual property rights are of much direct interest to the biotechnologist.

registered designs
Registered Designs protect the appearance of articles, insofar as this is not dictated by function. They have no relevance to life-forms, though it has been light-heartedly suggested that they might be available for plants of novel shape or colour. The shape is defined by photographs or drawings, and the method of making the design or the material from which it is constructed is not stated - so a suitable photograph of a blue rose might protect the rose whether made by plastic moulding or genetic engineering. Don't rely on it!

plant patents
The USA has "**Plant Patents**", introduced in 1930. These are in fact very similar to design registrations for plants. They are available only for asexually reproduced

non-tuber-propagated plants, and have been used mainly for ornamental plants such as roses and chrysanthemums. They have a specification with a description and drawing or photograph (usually in colour) and a claim which typically simply refers to the drawing.

Petty Patents (also known as Gebrauchsmuster) are available in some countries (Germany, Australia, for example). They are like patents, but with a shorter term (for example, six years) and a lower standard of inventiveness. Usually they are available only for devices, rather than substances, and are thus of limited interest for one who wishes to protect a process or an organism.

Petty Patents (Gebrauchs-muster)

copyright

"Copyright" is available for literary and artistic works. It relates to the form in which an idea is expressed, rather than the idea itself. Only copying is forbidden, which usually implies multiple copies - there are exceptions for research and private study. There is of course copyright in scientific papers, drawings and diagrams. This naturally includes gene sequences. It has been suggested that this should be a method for protection of the use of such sequences. However, unless the law was adapted substantially this would not work. Describing a material in a particular way (as by giving a gene sequence) does not give you special rights over that material. Copyright is in the original work-which is not the gene, but the writing down of the base sequence of the gene. And in any event, copyright could give no protection against someone who sequenced the gene independently, which would greatly restrict its usefulness.

Trade marks are a method of protecting reputation rather than technology. Nonetheless they can be invaluable to any businessman, if correctly handled. They certify the origin rather than the nature of goods. Just as a patentee may not monopolise methods or products that are old, so the trade mark owner may not reserve what other traders may reasonably wish to use. A trade mark should therefore not be descriptive of the goods sold. Moreover, if it becomes descriptive, rights in it may be lost. "Aspirin" was once a trade-mark - in a few countries, it still is. A catchy trademark is invaluable for a new product, but a common name is still necessary as well: if a new article becomes known only by its trademark, rights in the mark will be lost. For this reason we should not talk about a Thermos or a Walkman, but a "Thermos" brand vacuum flask and a "Walkman" brand personal stereo. This logic is worked out with plant varieties - the registered name of the variety is not a trade mark, but a description, and is free for all to use when the rights in the variety expire.

registered marks

Trade mark rights may be acquired by use, or by registration following use or (in many countries but not the USA) by registration with the intention of using. Registered marks last indefinitely, provided that renewal fees are paid, and that the mark does not lose its distinctiveness or becomes deceptive.

9.14 Intellectual property strategy for biotechnologists

Before starting a project:

1)   Do a patent and literature search. It is essential to know what others have done. You may find that what you wanted to do has already been done. Some areas may be patent-free: in others there will already be broad patent coverage. You may find technical disclosures that will help you carry out your work.

2)   Plan your research in the light of the results you obtain. If you are working in an area covered by a patent, will you be able to get a licence easily - or at all? If your

research is successful, will you be able to generate new patent property? If not, are there other ways in which you will be able to capitalise on your discoveries?

3) Consider filing patent applications before the work starts. You may have original ideas which you hope to verify. If so, file on them now. Patents are not awarded for verifying that an idea works, but rather for showing how to put an idea into effect. If you wait to confirm that your idea works, you run a considerable risk of having to take a licence from (or even being blocked by) someone else who filed first and proved it worked afterwards. However, this is a matter of fine judgement. You run the risk of having a patent application with a description which cannot be made to work.

During the research, you must keep up with the patent literature (by study of, for example, the EPO and US patent gazettes, and abstracting services such as Derwent).

If a competitor's patent is published on what you are doing, consider:

- will it be valid?

- should you oppose?

- would they license you?

- might they need a cross-licence from you?

As soon as you get results, consider adding to your existing patent applications and filing new ones.

Before filing abroad, ask:

- is your invention new and worthwhile?

- if you have enough data to support your patent (if not, can you risk abandoning your priority date and refiling)?

- in what countries do you need protection?

- which of those countries will give you useful protection?

- is the cost is worthwhile?

- is it an invention better kept as a secret?

9.14.1 Patent disputes

Avoid these if at all possible.

Litigation is always expensive. Patent litigation is exceptionally expensive, and the results are unpredictable. Moreover, litigation will take up inordinate amounts of time that would be better spent doing something else. It is better to take a license (or to grant one) under a patent of doubtful validity than to pay lawyers to discover whether the patent is valid or not. The penalty for being wrong is too high. Therefore, negotiate whenever reasonable. Of course, in games of incomplete information a mixed strategy is optimal. Sometimes it is necessary to sue (or be sued) so as to be able to present a credible threat in other circumstances. But not often.

9.15 Current developments

A number of changes to existing legislation are being discussed at present. Some of these could bring about significant changes in intellectual property protection.

Patent harmonisation

A patent harmonisation treaty has been under discussion for some years. It is being held up by similar negotiations in GATT (General Agreement on Tariffs and Trade, see below). The main changes being considered (other than those covered in GATT) are:

- USA adopts the first-to-file rule (abandoning the complexities of interference practice - see above);

- other countries adopt a grace period, in which the inventor can publish his invention without invalidating a subsequent application.

It is not yet clear whether either will go ahead.

European Community

Two significant items of legislation have recently (October 1992) passed the European Parliament. Their final form will be determined in the Council of Ministers.

The EC draft directive on patenting Biotechnology inventions has the objective of removing uncertainties in application of patent law to biotechnology. It is not intended to change existing conventions such as the European Patent Convention. Its passage through the European Parliament raised controversy on patenting 'life'. The current draft allows patenting lifeforms (in accordance with the EPC) but has introduced certain 'ethical' restrictions. Human beings and "human parts" are not patentable: nor are (inter alia) "unnaturally interspecific" animals.

The effects of these provisions are not clear. Is a human gene a "human part"? What about an animal (or totally synthetic) gene having homology (partial or complete) to a human gene? What about other "human parts"? Why should not a new form of mechanical heart valve be patented? What does "unnatural" mean? Would it cover any insertion into an animal of a gene from another species?

Another clause gives farmers an exemption from patent infringement for activities carried out on their own farms. The European Parliament has added a Farmer's Privilege which will discourage investment, if it survives.It remains to be seen how many of these provisions will survive into the final agreed directive.

The Draft Regulation on Plant Variety Rights proposes a unitary Plant Variety Right for the whole of Europe. The document emerging from the European parliament seems on the whole satisfactory, but some questions have been raised as to whether it is consistent with other new legislation, such as the draft Directive on patenting Biotechnology insertions described above and the revised UPOV described below.

Union for the Protection of New Varieties of Plants (UPOV) Convention

The latest version of the UPOV Convention was concluded in March 1991. It includes a number of important innovations, which have yet to be passed into law by any country.

Major changes included:

- stronger rights for breeders;
- restriction of farmer's rights;
- introduction of dependency.

Breeders' rights have been extended so that they are no longer necessarily limited to the right to multiply seed: they may cover harvested material and prevent importation of seed of the variety. They may even (at the option of each member country) control use of products directly derived from harvested material (flour, oil, etc).

Farmers' rights are codified. The norm is that such rights are not recognised: but since this is politically impractical in many countries at present, countries are allowed to provide exceptions, under specified conditions. The right must be confined to re-use by the farmer of seed produced by him on his own land; and the "legitimate interests of the breeder" must be safeguarded - which is taken to mean that the breeder must get some benefit (probably a reduced royalty).

Dependency is a new concept for plant breeding. The new convention provides that where a new distinct variety is derived predominantly from a protected variety, and is genetically closely similar, it will infringe the protected variety. Though a PVP will be granted for it, it can only be exploited with permission of the holder of the first variety.

The objects of this change are two-fold: to provide better protection for creative (as opposed to imitative) breeders; and in particular to protect breeders against molecular biologists. Up to now, anyone who inserted a single gene into a protected variety could thereby produce a distinct new variety which could be independently protected. There are some problems to be worked through in the application of this new concept (how similar is similar?) but in general it has been welcomed by innovative breeders. Without it, co-operation between breeders and plant biotechnologists would be handicapped by the poor bargaining position of the breeders.

GATT - General Agreement on Tariffs and Trade

At the time of writing (December 1992), prospects for completing the Uruguay Round of the GATT negotiations look more promising than for some time. One feature of these negotiations is TRIPs - Trade Related Intellectual Property. This aims to set minimum standards in intellectual property law in member countries. The latest draft (now unlikely to be changed much) provides for considerable strengthening of patent law. A much wider range of subject-matter must be patented - in principle all technology is to be treated alike, though exceptions permit countries not to grant patents on plants or animals (not just plant and animal varieties, as in European law). Inventions involving micro-organisms must be protected - thereby allowing biotechnological patents, presently not allowed in many countries. For countries who need to expand the protection they provide, a transition period is provided. There are also requirements to ensure that member countries have a suitable system of patent enforcement, and can grant interim injunctions in appropriate circumstances. Compulsory licensing of patents is also made subject to conditions.

TRIPS

Applying for market authorisation for medicinal products

Applying for market authorisation for medicinal products

10.1 General introduction

The commercialisation of medicinal products derived from biotechnology is a very complex process and is subject to a very wide range of Regulations. Some are specific for biotechnology, some are more general. Such regulations govern not only the scientific aspects but also the health and safety of workers, intellectual property rights, the legal issues of clinical testing, the licensing of medicinal products, liability and trademarks. This text does not purport to be an encyclopedia of all such regulations but to provide a reference source of the issues which particularly stem from biotechnology. Nor does this text claim to provide a training ground for medical lawyers. We have, therefore, been somewhat selective in our discussions to provide material of interest and value to biotechnologists in general.

At the end of this chapter, we provide a list of suitable references which give a more extensive treatment of the legislation and practices relating to the commercialisation of medicinal compounds.

10.2 Introduction to market authorisation

Within the EC, before a medicine can be introduced onto the market, the manufacturer must gain market authorisation. Gaining such authorisation requires submission of substantive documentary material to a "competent authority". Such documentary evidence covers such issues as:

- the clinical use of the product;

- the method of manufacture;

- the quality control measures to be undertaken;

- the chemical and physical properties of the product;

- the results of pharmacological tests;

- the results of toxicity tests;

- the results of clinical studies.

The contents of such a dossier are complex and must respond to a series of EC-Directives and Guidelines. The submission and evaluation processes are also complex.

In this chapter we will first outline to whom Market Authorisation dossiers should be submitted and briefly list the documents (Directive etc) which need to be complied with. Subsequently we will examine the application process itself. In the next chapter we will consider the principles of Good Manufacturing Practices (GMP) and their relationship with Good laboratory Practices (GLP), Quality Assurance (QA) and Quality Control (QC).

We draw your attention to the fact that the EC's Committee for Proprietary Medical Products (CPMP), in consultation with the competent authorities of the Member States, has produced a series of Notices for Applicants under the heading "The Rules governing Medicinal Products in the European Community".

This is in the form of five volumes as outlined below:

Volume I	The rules governing medicinal products for human use in the European Community
Volume II	Notice to applicants for marketing authorizations for medicinal products for human use in the member states of the European Community
Volume III	Guidelines on the Quality, Safety and Efficacy of medicinal products for human use
Volume IV	Guide to Good Manufacturing Practice for the manufacture of medicinal products
Volume V	The rules governing medicinal products for veterinary use in the European Community

These Notices for Applications are of course, subject to updating and it is important that these 'Notices' or their replacements are referred to before making an application for Market Authorisation. In this chapter we will particularly refer to the issues concerning making an application (volume II) and in the next chapter we will discuss Good Manufacturing Practises (volume IV).

10.3 Competent authorities and market authorisation (MA)

national
authorities
supernational
authorities

Historically, in Europe, MA files were submitted to national authorities. Each national authority would apply national regulations in making a judgement on each application. Thus a product would need to be submitted for evaluation in several states if the intention was to manufacture or use the product in those states. The emergence of the European Community has led to several modifications in the procedure and a manufacturer now has some choice between submitting licensing files to national authorities or to supranational authorities such as the EC Committee for Proprietary Medicinal Preparations (CPMP).

multi-state
application
supranational
body

Since 1977, as far as proprietary medicinal products are concerned, it has been possible to use such a supranational body to obtain marketing authorisation in EC member states. Instead of submitting individual applications, a common application to five (reduced to two in 1985, 83/570/EEC) or more member states is made using the CPMP. This, however, can only be done after authorisation has been obtained in one member state. This first application is crucial because the decision, assuming it is favourable, must be taken into account by other Member States as they become involved.

The sequence is therefore:-

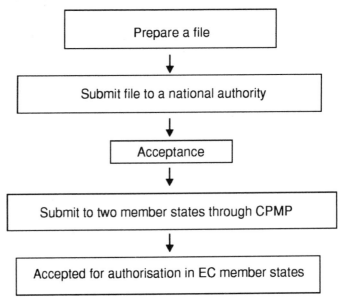

In 1987 the above, multi-state procedure was extended to 'high tech' products (ie those obtained by genetic engineering and monoclonal antibodies) even when the intention was to submit to a single member state (87/SS/EEC).

The objective was to enable questions relating to the quality, safety and efficiency of biotechnological medicinal products to be resolved within the EC by CPMP. This procedure would facilitate subsequent access to the markets of other member states and enable a co-ordinated and consistent stance to be taken within the EC member states. A list of addresses of the 'Competent Authorities' to whom application may be made is provided in Table 10.1.

Table 10. 1

Names and addresses of competent authorities

Belgium:
Ministere de la Sante publique
Insepction generale de la
Pharmacie
Cite Administrative, Quartier
Vesale
B-1010 Bruxells
Ministerie van Volksgezondheid
Farmaceutishe Inspektie
Vesalius Gebouw, B-1010 Brussel
Tel: (32) (2) 210 4900 and 210 49 01
Telex: 25768 MVGSPF B
Telefax (32) (2) 210 48 80

Denmark:
Sundhedsstyrelsen
Lagemiddelafdelingen
Frederikssundsvej 378
DK-2700 Bronshoj
DK-2700 Bronshoj
Tel: (45) (2) 94 36 77
Telex: 35333 IPHARM DK
Telefax: (45) (2) 84 70 77

continued

Spain:

99 Ministeria de Sanidad y
Consumo Direccion General de
Farmacia y
Productos Sanitarios
Paseo del prado, 18-20
E-28014 Madrid
Tel: (34) (1) 467 34 28, Telex:
22608 MSASS Telefax: (34) (1)

Germany:

Institute fur Arzneimittel des
Bundesgesundehitsamtes
Seestr 10
D-1000 Berlin 65
Tel: (49) (30) 45
Telex: (2627) (17) 308062 BGESA D
Felefax: (49) (30) 4502207

For sera, vaccines and allergens:

Paul-Enrlich-Insitut
Bundesamt fur Sera and Impfstoffe
Paul-Ehrlich-Str 42-44
Postfach 700810
D-6000 Frankfurt/M 70
Tel: (49) (69) 634402
Telex/Teletex: 6990716
Telefax: (49) (69) 63 44 02

Italy:

Ministero della Sanita, Servizio
Farmaceutico Viale della Civilta
Romana, 7
1,00144 Roma, EUR
Tel: (39) (6) 592 58 63
Telex: 625205 MINSANI
Telefax: (39) (6) 592 58 24

Netherlands:

College ter beoordeling van
geneesmiddelen P.O. Box 5811
NL-2280 HV Rijswijk
Tel: (31) (70) 40 70 08
Telex: 31680 WVCRW NL
Telefax: (31) (70) 40 50 48

France

Ministere de la Solidarite
de la Sante it de la Protection sociale
Direction de la Pharmacie de du
Medicament 1, place de Fontenoy
F-75700 Paris
Tel: (30) (1) 323 0911, Telex: 223514
Telex: 250011 SANTSEC F
Telefax: (30) (1) 323 86 81

Greece:

E.O.F. (National Drug
Organisation)
Voulis Str 4
Athens 10562
Tel: (30) (1) 323 0911, Telex: 223514
Telefax: (30) (1) 323 8681

Ireland:

National Drugs Advisory Board
63-64 Adelaide Road
Dublin 2
Tel: (353) (1) 76 4971 - 7, Telex:
90542 Telefax: (353) (1) 78 60 74

Luxembourg:

Direction de la Sante
Division de la Pharmacie et des
Medicaments 10, rue C.M. Spoo
L-2546 Luxembourg
Tel: (352) 4 08 01, Telex: 2546 SANTE
LU

Portugal:

Ministeria da Saude
Direccao Geral dos Assuntos
Farmaceuticos Av Estados Unidos
da America, 37
P-1700 Lisboa
Tel: (351) (1) 80 41 31
Telex: 15655 MADP
Telefax: (351) (1) 88 03 31

continued

United Kingdom:

Department of Health and Social
Security Medicines Division
Market Towers
1 ine Elms Lane
London SW8 5NQ
Tel: (44) (1) 720 2188
Telex: 883669 DHSSHQ G
Telefax: (44) (1) 720 5647

For delivery of the dossier:

Department of Health
Medicines Division
Britannia House
7 Trinity Street
London SE1 1DA
Tel: (44) (1) 407 5522

Address of the Secretariat of the
Committee for Proprietary
Medicinal Products
DGGIII B 6 "Pharmaceuticals,
veterinary medicines", Commission
of the European Communities, rue
de la Loi 200, B-1049 Brussels,
Telephone: 236 03 32/235 69 35,
Telex: 21877 COMEU B

USA and the FDA

In the USA the administrating body for the introduction of new medicines is the Food and Drug Administration (FDA). In the USA, the Codes of Federal Regulation (CRF) describes both administrative and specific requirements. The specific requirements are confined to drugs that have existed for at least a decade (eg vaccines, sera and blood derived drugs). For more recent products, like monoclonal antibodies and those involving recombinant DNA, so called 'points to consider' have been drafted. There is particular concern to stimulate discussion within industry.

Japan and the PAB

In Japan, the body responsible for reviewing applications to licence new drugs is the Pharmaceutical Affairs Bureau (PAB) of the Ministry of Health and Welfare. In the main, the EC, FDA and PAB are attempting to harmonise their approach to licensing medicines. This is especially true in the approach to the licensing of medicines derived from biotechnology. It is not surprising that other licensing authorities are following the lead of the EC, FDA and PAB. Thus, although some important differences still exist, there is a growing trend towards a global approach.

In order to maintain clarity we will focus on the situation in Europe.

10.4 What documents need to be complied with?

EC directive and guidance documents

The area of market authorisation for medical products has been particularly subject to a sequence of new EC Directives and guidance documents. In the evolving European situation, many Directives and guidance notes have been issued. We can expect there will be many more. A substantial, but not complete list is provided here. It is our intention to explore the structure of the regulations and to provide some information on the EC approach to market authorisation. The documents that need to be complied with to gain market authorisation, fall into two general categories: broad ranging EC Directives; and EC Guidelines. The reader should note that EC Directives do not bind the members of the European Community, but contain an obligation for the Member States to bring into force laws, regulations and administrative provisions necessary to comply with the directive concerned. Guidelines on the other hand provide help to applicants to provide the information that is required before authorisation can be given.

A list of the major community Directives is given in Table 10.2. Note that these documents are published in the "Official Journal" (OJ) which is available in all Member States. Copies of the Official Journal are usually placed in main libraries or can be obtained from the offices of the Official Journal in each Member State.

A list of Guidelines is given in Table 10.3. Some further relevant EC communications are given in Table 10.4.

EC Directives   OJ = Official Journal

- Council Directive 65/65/EEC of 26 January 1965 on the approximation of provisions laid down by law, regulation or administrative action relating to proprietary medicinal products (O. J. no L22 of 9.2.65).

- Council Directive 75/318/EEC of 20 May 1975 on the approximation of the laws of Member States relating to analytical, pharmaco-toxicological and clinical standards and protocols in respect of the testing of proprietary medicinal products (O. J. No L 147 of 9.6.75).

- Council Directive 75/319/EEC of 20 May 1975 on the approximation of provisions laid down by law, regulations or administrative action relating to proprietary medicinal products.

- Council Decision 75/320/EEC of 20 May 1975 setting up a Pharmaceutical Committee (O. J. No L 147 of 9.6.75)

- Council Directive 78/25/EEC of 12 December 1977 on the approximation of the laws of the Member States relating to the colouring matters which may be added to medicinal products (O. J. no L 11 of 14.1.78).

- Council Directive 81/851/EEC of 28 September 1981 on the approximation of the laws of the Member States relating to veterinary medicinal products (O. J. no L 317 of 6.11.81).

- Council Directive 81/852/EEC of 28 September 1981 on the approximation of the laws of the ember States relating to analytical, phamacotoxicological and clinical standards and protocols in respect of the testing of veterinary medicinal products (O. J. no L 317 of 6.11.810.

- Commission Communication on parallel imports of proprietary medicinal products for which marketing authorizations have already been granted (O. J. no C 115 of 6.5.82).

- Council Directive 83/189EEC of 28 March 1983 laying down a procedure for the provision of information in the field of technical standards and regulations (O. J. no L 109 of 26.4.83).

- Council Directive 83/570/EEC of 26 October 1983 amending Directives 65/65/EEC, 75/318/EEC and 75/319/EEC on the approximation of provisions laid down by law, regulations or administrative action relating to proprietary medicinal products (O. J. no L 332 of 28.11.83).

- Council Recommendation 83/571/EEC of 26 October concerning tests relating to the placing on the market of proprietary medicinal products (O. J. no L 332 of 28.11.83).

(Continued)

Table 10.2 A library of EC Directives relating to medicinal products.

- Commission Communication on the compatibility with Article 3d of the EEC Treaty of measures taken by Member States reacting to price controls and reimbursement of medicinal products (O. J. no C 310 of 4.12.86).

- Council Directive 86/609/EEC of 24 November 1986 on the approximation of laws, regulations and administrative provisions of the Member States regarding the protection of animals used for experimental and other scientific purposes (O. J. No L 358 of 18.12.86).

- Council Directive 87/18/EEC of 18 December 1986 on the harmonisation of laws, regulations or administrative provisions relating to the application of the principles of good laboratory practice and the verification of their applications for tests on chemical substances (O.J.No L 15 of L15.

- Council Directive 87/19/EEC of 22 December 1986 amending Directive 75/318/EEC on the approximation of the laws of the Member States relating to analytical, pharmaco-toxicological and clinical standards and proposals in respect of the testing of proprietary medicinal products (O. J. No L 15 of 17.1.87).

- Council Directive 87/20/EEC of 22 December 1986 amending Directive 81/852/EEC on the approximation of the laws of the Member States relating to analytical, pharmaco-toxicological and clinical standards and protocols in respect of the testing of veterinary medicinal products (O. J. no L 15 of 17.1.87).

- Council Directive 87/21/EEC of 22 December 1986 amending Directive 65/65/EEC on the approximation of provisions laid down by law, regulations or administrative action relating to proprietary medicinal products (O. J. no L 15 of 17.1.87).

- Council Recommendation 87/176/EEC of 9 February 1987 concerning tests relating to the placing on the market of proprietary medicinal products (O. J. no L 73 of 16.3.87).

- Council Directive 88/182/EEC of 22 March 1988 amending Directive 83/189/EEC laying down a procedure for the provision of information in the field of technical standards and regulations (O.J. no L 51.26.3.88).

- Council Directive 88/320/EEC of 9 June 1988 on the inspections and verification of Good Laboratory Practice (GLP) (O. J. No L 145 of 18.12.86).

- Council Directive 89/105/EEC - Council Directive relating to the transparency of measures regulating the pricing of medicinal products for human use and their inclusion within the scope of national health insurance systems

- Council Directive 89/341/EEC - Council Directive amending Directives 65/65/EEC, 75/318/EEC and 75/319/EEC on the approximation of provisions laid down by law, regulation or administrative action relating to proprietary medicine

- Council Directive 89/342/EEC - Council Directive extending the scope of Directives 65/65/EEC and 75/319/EEC and laying down additional provisions for immunological medicinal products consisting of vaccines, toxins or serums and allergens

- Council Directive 89/343/EEC - Council Directive extending the scope of Directives 65/65/EEC and 75/319/EEC and laying down additional provisions for radiopharmaceuticals

- Council Directives 89/381/EEC - Council Directive extending the scope of Directives 65/65/EEC and 75/319/EEC and laying down additional provisions for medicinal products derived from human blood or human plasma

Table 10.2 A library of EC Directives relating medicinal products.

EEC Guidelines

a) Recommendation 83/571/EEC relates to the following five notesfor guidance, published in the O. J. no L 332 of 28.11.83: repeated dose toxicity; reproduction studies; carcinogenic potential; pharmaco-kinetics and metabolic studies in the safety evaluation of new drugs in animals; fixed-combination products.

b) Recommendation 87/176/EEC includes the following fourteen notes for guidance, published in the O. J. no L 73 of 16.3.87: single dose toxicity; testing of medicinal products for their mutagenic potential; cardiac glycosides; clinical investigation of oral contraceptives; user information on oral contraceptives; data sheets for antimicrobial drugs; clinical testing requirements for drugs for long-term use; non-steroidal anti-inflammatory compounds for the treatment of chronicdisorders; anti-epileptic/anticonvulsant drugs; investigation of bio-availability; clinical investigation of drugs for the treatment of chronic peripheral arterial diseases; pharmaco-kinetic studies in man; anti-anginal drugs; topical corticosteroids.

c) Other guidelines adopted or in preparation by CPMP.

- recommended basis for the conduct of clinical trials of medicinal products in the European Community:

- production and quality control of monoclonal antibodies of murine origin intended for use in man;

- production and quality control of medicinal products derived by recombinant DNA technology;

- chemistry of the active ingredient;

- stability tests on active substances and finished, products;

- development pharmaceutics ad process validation;

- preclinical biological safety testing; antidepressant drugs;

- testing medicinal products in the elderly; clinical trials in children;

- herbal remedies; trials of medicinal products in the treatment of cardiac failure; antiarrhythmic drugs;

- anti-cancer agents in man, III/699/88;

- studies of prolonged action forms in man, III/1962/87;

- production and quality control of cytokine products derived by modern biotechnological processes,

- III/3791/88; anti-epileptic/anticonvulsant drugs, II/3128/88: revision; analytical validation,

- III/844/87; general phamacodynamics,

- III/480/87; production and quality control of monoclonal antibodies derived from human lymphocytes intended for use in man, III/3975/88;

- good clinical practices, III/3976/88;

- local toxicity and skin/eye toxicity, III/3979/88;

- control tests in the finished product, III/3978/88;

- mental deficiency in the aged, III/3977/88.

Table 10.3 Community guidelines prepared with the Committee for Proprietary Medicinal Products.

EC Communications

COM (90) 101 final - SYN 255 - Proposal for a Council Regulation concerning the creation of a supplementary protection certificate for medicinal products

COM (88)496 final - Proposal for a Council Directive on the legal protection of biotechnological inventions

COM (89) 607 final - SYN 229 - Proposal for a Council Regulation concerning the creation of a supplementary protection certificate for medicinal products

COM (88) 496 final - Proposal for a Council Directive on the legal protection of biotechnological inventions

COM (89)607 final - SYN 229 - Proposal for a Council Directive on the wholesale distribution of medicinal products for human use

COM (89)607 final - SYN 230 - Proposal for a Council Directive concerning the legal status for the supply of medicinal products for human use

COM (89)607 final - SYN 231 - Proposal for a Council Directive on the labelling of medicinal products for human use and on package leaflets

COM (90) 212 final - SYN 273 - Proposal for a Council Directive on advertising of medicinal products for human use

COM(90)283 final - SYN 309 - Proposal for a Council Regulation (EEC) laying down Community Procedures for the authorization and supervision of medicinal products for human and veterinary use and establishing a European Agency for the Evaluation of medicinal products

Table 10.4 Some relevant communications of the EC.

10.5 The Key EC-Directives dealing with application procedures for market authorisation

EC-Directive 65/65	- on the approximation of provisions laid down by law, regulation or administrative action relating to proprietary medicinal products (OJ No L22)
EC-Directive 75/319	- on the approximation of provisions laid down by law, regulation or administrative action relating the proprietary medical products (OJ NoL147)
EC-Directive 83/570	- amending EC-Directives 65/65/EEC, 75/318/EEC, 75/319/EC on approximation of provisions laid down by law, regulation or administrative action relating to proprietary medical products (OJ NoL 332)
EC-Directive 87/22	- on the approximation of national measures relating to the placing on the market of high technology medicinal aproducts, particularly those derived from biotechnology. (OJ No L15)

Central to the application for market authorisation is Article 4 of Directive 65/65/EEC which demands that a dossier (file) be compiled for submission to the relevant competent authority.

In the following section we outline the contents of the EC-Directives concerning the procedures for making a multi-state application for authorisation (Directive 83/570/EEC) and the concertation procedure for the marketing of high technology products, particularly those derived from Biotechnology (Directive 87/22/EEC).

10.6 The multi-state procedure (Directive 83/570/EEC)

10.6.1 Purpose and scope of the multi-state procedure

The rules governing the "multi-state" procedure are set out in Chapter III of Directive 75/319/EEC, as amended by Directive 83/570/EEC. Particulars and documents supporting the application must, if necessary, be updated to comply with the rules in force at the time of submission.

The primary purpose of the multi-state procedure is to make it easier for a person who has already obtained a MA in one Member State to get further MAs for the product concerned in two or more of the other Member States. On the basis of the same complete documentation, and taking the MA granted by the first Member State into due consideration, the authorities of the Member States to which the application is addressed have 120 days to grant authorisation to market the product in their country or to formulate reasoned objections. Where one or more objections are advanced, the matter is referred to the CPMP which considers the grounds for the objections and any written or oral explanations provided by the applicant before issuing its own reasoned opinion, normally within a period of 60 days. This opinion is addressed to the Member States concerned, and is also communicated to the applicant. Within a further 60 days the Member States must decide on what action to take pursuant to the Committee's opinion and must inform the CPMP of their decision.

In addition, the multi-state application must relate to a proprietary medicinal product which has been authorised by one Member State in accordance with the criteria laid down by the Community directives. Thus products marketed by virtue of previous national provisions are not covered by the new procedures unless their quality, safety and efficacy have been reviewed in accordance with Article 39 of Directive 75/319/EEC.

10.6.2 New drug applications and abridged applications

The multi-state procedure may be used not only for new drug applications but also for abridged applications submitted pursuant to point 8 of Article 4 of Directive 65/65/EEC. The rules governing the submission of abridged applications have been amended by Council Directive 87/21/EEC which entered into force on 1 July 1987. These new rules apply both to applications submitted in accordance with the multi-state procedure and applications submitted in accordance with national procedures.

requirement for
a
comprehensive
dossier and
expert reports

A complete chemical, pharmaceutical and biological dossier will always be required together with 3 Expert Reports. An applicant will not be required to submit the results of pharmacotoxicological tests and clinical trials if he can demonstrate that the proprietary medicinal product for which application is made is essentially similar to another product which has been authorised within the Community in accordance with Community provision in force for a specified period and is marketed in the Member State or Member States to which application is made. However, in such cases, important references should be supplied in full with the application. As appropriate, the expert reports may be abbreviated, but the essential similarity must be critically evaluated. In the case of multi-state applications, the "similar product" must be marketed in all the Member States concerned by the application. For the purpose of this provision, a proprietary medicinal product will be regarded as essentially similar to another product if it has the same qualitative and quantitative composition in terms of active principles and the pharmaceutical form is the same and where necessary, bio-equivalence with the first product has been demonstrated by appropriate bio-availability studies carried out, in accordance with the principles set out in Annex X to Council Recommendation 87/176/EEC of 9.2.87 (see also the Annex to Directive 75/318/EEC). These factors should be fully discussed in the relevant Expert Reports.

The specified period after which the applicant is no longer required to provide the results of pharmaco-toxicological tests and clinical trials is:

- in the case of proprietary medicinal products which have been derived from biotechnology, and other high technology medicinal products on which the CPMP has given an opinion, ten years, from the date of the first national MA was granted within the Community;

- in the case of other proprietary medicinal products, six years, from the date of the first national MA within the Community. However this six year period may be extended to ten years by Member States.

It should be noted that these periods do not prejudice the patent rights of the manufacturer of the original product. Any application for a MA submitted before the expiry of the periods referred to must be complete except in cases where:

- the proprietary medicinal product is essentially similar to a product authorized in the country concerned by the application and the person responsible for the marketing of the original proprietary medicinal product has consented to a specific subsequent applicant referring to the information contained in the dossier;

- by detailed references to published scientific literature presented in accordance with the second paragraph of Article 1 of Directive 75/3187/EEC, it is demonstrated that the constituent or constituents of the proprietary medicinal product have a well established medicinal use, with recognised efficacy and acceptable level of safety. In such cases the full article or reference should be supplied (with necessary translations), and, as appropriate, individual study reports and clinical formats should be prepared. Published scientific literature includes journals, books and articles in the public domain, which are generally available. Moreover the Expert Reports must clearly state the grounds for using published references under the conditions set out in Directive 75/319/EEC;

- detailed reference to published literature (as above) is supplemented with appropriate additional studies.

However, where the proprietary medicinal product is intended for a different therapeutic use from that of the other proprietary medicinal products marketed or is to be administered by different routes or in different doses or in different dosage forms, the result of appropriate pharmaco-toxicological tests and/or of appropriate clinical trials may need to be provided. In the case of new proprietary medicinal products containing known constituents not hitherto used in combination for therapeutic purposes, the results of pharmaco-toxicological tests and of clinical trials relating to that combination should be provided, and be discussed in the Expert Reports. It is not, however, necessary to provide references relating to each individual constituent. The fact that in one Member State a given constituent is well-known does not necessarily imply that the same is true of all the other Member States. In certain Member States there may be no experience of the medicinal use of the constituent and complete documentation may, therefore, be required.

10.6.3 Submission of the multi-state application

steps in the
submission of
the application

In the first instance, the applicant should consult the competent authority of the Member State which grants the initial authorisation about any additions which have been made to the original dossier to bring it up-to-date and obtain approval for such additions. This authority may require the applicant to provide such information and documents as are necessary to enable it to check the identity of the dossier filed under the multi-state procedure with the dossier upon which it took its own decision.

The person responsible for placing the product on the market submits an application directly to the competent authorities of each of the two or more Member States concerned, referring to the procedure laid down in Chapter III of Directive 75/319/EEC, as amended by Directive 83/570/EEC.

It is considered inappropriate to submit an application in some Member States using the multi-state procedure, while, simultaneously in other Member States the same application is submitted through national procedures. Applicants are, therefore, urged to involve all Member States concerned with the application in the multi-state procedure. Each application should be accompanied by the documents referred to in Articles 4, 4(a) and 4(b) of Directive 65/65/EEC.

The applicant should also testify that each dossier is identical to that accepted by the first Member State, or when necessary, he should specify any additions or modifications that have been made.

It is recommended that the same trade name for a given medicinal product should be used throughout the Community, unless adequately justified.

In addition, it should be indicated whether an application to market the proprietary medicinal product has been made to, or granted by, any other Member State under purely national procedures.

The applicant notifies the Secretariat of the CPMP of the multi-state application, informs it of the Member States concerned and of the dates on which the dossiers were sent to those Member States and sends it a copy of the authorisation, including the summary of the product characteristics approved by the first Member State. The applicant also sends to the Secretariat a copy of the authorisation, including the summary of the product characteristics approved by the first Member State (Article 4(b) of Directive 65/65/EEC).

After all the Member States concerned have confirmed receipt of the application, the Secretariat notifies all the Member States and the applicant of the start of the 120 day period referred to in Article 9 (3) of Directive 75/319/EEC (as amended).

National registration fees remain payable in accordance with the rules of the Member State concerned.

role of
competent
authority in
communicating
with other
Member States

In accordance with Article 13 of Directive 75/319/EEC as amended, immediately a multi-state application is lodged, the competent authority of the Member State which granted the initial application is required to communicate a copy of any assessment report relating to the particular product to the Member States concerned by the application. Persons preparing a multi-state application are, therefore, strongly advised to contact this authority at an early stage to make the administrative arrangements necessary to ensure that any necessary translations of the assessment report into a language or languages acceptable to the countries concerned by the application, are available at the same time as they formally submit the application. The applicant will be expected to pay for any translations of the assessment report which are necessary to consider his application. In order to reduce these costs to a minimum, all the competent authorities have indicated that they are prepared to accept official assessment reports in English if they are not available in their own national language.

The multi-state procedure is started by the submission of an application by the holder of the first authorisation. For practical reasons, co-ordination of objections and a single response may be made through a nominated (central) office of the company. Companies who are to be the holders of authorisations granted under the procedure should be stated at the outset.

10.6.4 Presentation of the multi-state Application

a) Order of presentation and content of the dossier

Detailed guidance on the presentation of the different parts of a multi-state application can be obtained from CPMP. This guidance has been prepared and accepted in consultation with the competent authorities of all the Member States, in order to facilitate the examination of the application. Clinical documentation, (case report forms) may be requested by Member States and should therefore always be available. They may be requested during the assessment or at the date of submission.

b) Expert Reports

In accordance with Directive 75/319/EEC, Article 2, the pharmaceutical, pharmaco-toxicological and clinical parts of the complete dossier should each include an Expert Report. It is important to emphasize that well prepared Expert Reports greatly facilitate the task of the competent authorities in evaluating the dossier and contribute towards the speedy processing of applications. For these reasons particular care should be taken in the preparation of the Experts Reports. Guidance can be obtained from CPMP.

Where relevant Community guidelines on the conduct of tests and trials on a proprietary medicinal product exist, these should be taken into consideration when the report is prepared and any deviation from them should be discussed and justified. A list of the guidelines currently available is included in Table 10.3.

c) Summaries of the dossier

Where, in addition to copies of the complete dossier, a Member State concerned by an application requires the submission of additional copies of summaries of the dossier, these should consist of at least the information contained in Part I of the application

"Summary of the Dossier": i.e. Administrative Data and the Summary of Product Characteristics (Article 4 (a) of Directive 65/65/EEC as amended) together with the Experts Reports and tabulated formats on the three parts of the dossier: pharmaceutical, pharmacotoxicological and clinical.

d) Number of copies and accepted languages

Information on the number of copies of the application to be submitted to the authorities of each Member State and on the languages in which applications should be drafted is set out in the Table 10.5. Further copies of the application may be required by the authorities in certain exceptional cases.

e) Specimens and Samples

In accordance with Article 4, second paragraph, point 9, of Directive 65/65/EEC as amended, a specimen or mock up of the sales presentation of the proprietary product, together with a package leaflet where one is to be enclosed, must be included in each complete dossier submitted. The summary of product characteristics, specimens or mock-ups of the sales presentation together with a package leaflet where one is to be used enclosed, should be translated into the national languages(s) of the Member State concerned by the application.

Moreover, for the purposes of implementing Article 4 of Directive 75/319/EEC, samples of the active principles and of the finished product must be supplied, as a matter of course, to the competent authorities in Belgium, Greece, Ireland, Italy, Luxembourg, Spain, the Netherlands and Portugal. In other cases, samples should be provided at the request of the competent authorities.

10.6.5 Cases where a CPMP opinion is required

As noted above, an opinion of the CPMP is not required if no Member State has put forward any reasoned objections during the 120 day period allowed for national examination of the application. In such a case, marketing authorisation will have been granted by each of the Member States concerned.

Table: 10.5 Number of copies and acceptable languages							
	BE	**DK**	**DE**	**GR**	**SP**	**FR**	**IR**
Number of Copies	i)				ii)	ii)	
Full dossiers	1	1	4	1	2	2	1
						v)	
Part II Pharmaceutical	+2	-	-	-	+3	+3	-
Additional summaries i.e.	5	1	10	5	12	50	3
Parts I and V; including Expert Reports							
Consolidated responses	2	2	4	2	2	2	2
II. Languages in principle	FR or NL	DK	DE	GR	SP	FR	EN
Other languages accepted for Part II(Pharmaceutical)	EN or DE	EN	EN	EN	SP	EN	EN
					iii)	iii)	
Part III (Pharmaco-toxicological) Part IV (Clinical)	EN or DE	EN	EN	EN	EN or FR	EN	EN
			iv)				
Summaries including Expert Reports	EN	EN	EN	EN	SP	FR	EN

i) Additional copies of Part IV (A) Clinical Pharmacology required - one for NL; two for BE only on request.

ii) Additional copies of Part II Toxicology required - one for FR; one for SP.

iii) For SP, PR, PO all studies or trials are accepted in the other language mentioned if accompanied, in the national language, by a very detailed and precise summary, particularly with references to the pages containing the data in question.

iv) The submission of these summaries in German is preferred.

Table 10.5 Number of copies and the language to be used for applications submitted to the authorities in Member States.   (Continued)

Table: 10.5 Number of copies and acceptable languages (Continued)						
	IT	LX	NL	PO	UK	CPMP
Number of Copies			i)			
Full dossiers	2	1	2 5)	2	3	1
Part II Pharmaceutical	-	+2	+1	+1	+1 7	-
Additional summaries i.e.	10	1	3	4	-	2
Parts I and V; including Expert Reports Consolidated responses	2	2	2	2	2	2
II. Languages in principle	IT	FR	NL	PO	EN	EN or FR
Other languages accepted for Part II	EN or FR	EN or DE	EN FR DE	EN or FR	EN	EN or FR
(Pharmaceutical)or FR	FR DE 3)	EN or DE 3)	EN or FR	EN FR	EN EN 3)	EN
Part III (Pharmaco-toxicological) Part IV (Clinical)	EN or FR	EN or FR	EN	EN	EN	EN
Summaries including Expert Reports	EN or FR	EN or DE	EN FR DE	PO or FR	EN	EN

i) Additional copies of Part IV (A) Clinical Pharmacology required - one for NL; two for BE only on request.

ii) Additional copies of Part II Toxicology required - one for FR; on for SP.

iii) For SP, PR, PO all studies or trials are accepted in the other language mentioned if accompanied, in the national language, by a very detailed and precise summary, particularly with references to the pages containing the data in question.

iv) The submission of these summaries in German is preferred.

Table 10.5 Number of copies and the language to be used for applications submitted to the authorities in Member States.

action when objections are lodged

If, however, one or more of the Member States concerned does lodge reasoned objections to the application within the 120 days, the objections are formally notified to the applicant by the authority concerned and the matter is referred to the CPMP for its opinion. In this case, the applicant is required to send to the Secretariat of the Committee a complete copy of the application together with 2 summaries of the dossier as soon as possible after the receipt of reasoned objections from a Member State. It is suggested, particularly in the case of large dossiers, that contact be made with the

Secretariat in advance, to arrange for receipt of the documents. In addition, a copy of the summary of the dossier should be sent directly to each of the competent authorities who are not directly concerned by the application. Although the complete dossier may be submitted in English or French, it will greatly facilitate the work of the Committee if copies of these summaries are made available in both English and French. At the same time copies of any already existing assessment reports will be circulated by the Member States, with a copy to the Secretariat.

The Committee is required to give its opinion within 60 days of the date on which the matter is referred to it. The 60 days allowed for examination of the application by the CPMP will begin once all those concerned have received the additional documentation. Applicants should liaise with the original Member State regarding:

- the objections raised;

- answers and information proposed in response to these objections;

- the time period required to prepare responses;

- whether a hearing is necessary, and if so, the nature and format of such hearings.

In addition, in accordance with Article 14(1) of Directive 75/319/EEC as amended, an applicant may provide the Committee with a written response and/or he may request a hearing to provide the Committee with oral explanations. As soon as possible after the receipt of reasoned objections from a Member State, the Secretariat of the Committee will notify the applicant of the date on which the Committee proposes to consider the application and of the deadline for the submission of any written response the applicant may wish to make in accordance with Article 14 of Directive 75/319/EEC, as amended. In the case where an oral hearing is requested before the Committee, this must be confirmed in writing 30 working days before the hearing. If the applicant considers that the date proposed for the examination of his application by the Committee does not provide sufficient time for the preparation of a response, he may request that examination of the application be postponed to a subsequent meeting. At the present time the Committee usually meets at least once every two months.

The reasoned opinion of the CPMP is exclusively concerned with the grounds for the objections put forward by the Member States concerned. The opinion of the Committee, or, in the case of divergent opinions, the opinions of its Members are immediately notified to the applicant and to the Member States.

CPMP opinion does not replace national decisions

The opinions of the CPMP do not replace national decisions. However, within 60 days of the receipt of the opinion, the Member States concerned must decide what action to take on the Committee's opinion and inform the Committee of that decision. The Member States keep the Committee informed of the action, in accordance with individual administrative procedures, they are taking pursuant to an opinion until such time as a definitive decision is adopted.

In accordance with Article 214 of the EEC Treaty and Article 19 of its rules of procedure, the deliberations of the CPMP and all documents submitted are confidential. Dossiers can be held in the Secretariat only as long as necessary, and normally will be available to the applicant once the opinion of the CPMP has been expressed. Applicants should, at the commencement of the procedure, make arrangements regarding the collection of dossiers, or alternatively, supply written confirmation that the dossier may be destroyed.

10.6.6 Written representations and oral hearings

The purpose of written representations and oral hearings is:

- in the case of multi-state applications for MA submitted pursuant to Article 9 of Directive 75/319/EEC as amended, to enable the applicant to provide explanations in response to the reasoned objections put forward by one or more of the Member States concerned by the application;

- in the case of matters referred to the Committee pursuant to Article 11 of Directive 75/319/EEC as amended, to enable, where the Committee considers it appropriate, the person responsible for marketing the product to provide explanations in response to the grounds for the refusal, withdrawal or suspension of the MA given by one or more Member States.

all Member States involved

Applicants using the multi-state procedure should bear in mind that oral or written explanations are made to the CPMP as a whole, which comprises representatives from all the Member States of the Community, and not just to the countries concerned by the application. Although the authorities which are not directly concerned by the application will not necessarily have seen the complete dossier, they will have seen the summaries of the dossier, the reasoned objections of the Member States which are directly concerned by the application and any available assessment reports.

In order to enable the CPMP to concentrate on the important issues raised concerning the acceptability of a proprietary product on grounds of quality, safety and efficacy, applicants using the multi-state procedure are advised to try to resolve any minor objections, including those concerning the chemical, pharmaceutical and biological part of the dossier, directly with the competent authorities concerned, if possible before the date on which the Committee will consider the application. However applicants should ensure that any additional information which is provided in response to such objections is circulated to all the Member States, to the Secretariat, and that , whenever necessary, the Expert Reports and summaries are updated. Whenever possible, this information should be circulated together as a single consolidated updating of the application. Any written response from the applicant should be sent directly to all the Members of the CPMP, with two copies to the Committee's Secretariat (and 10 copies of relevant text for use by interpreters, and should reach the Members at least 30 working days before the meeting, otherwise consideration of the application will need to be postponed. The response should include all objections raised by concerned Member States, with full explanation for each, noting the origin of each objection and should be presented in the same format as the dossier. Sufficient numbers of copies of the response should be supplied to each member of the Committee to allow for internal consultation in the competent authority. Any written response should set out the name of the proprietary product concerned, its composition in terms of active principles and the name and address of the person responsible for marketing the product.

Although the CPMP does not, at present, wish to lay down formal rules of procedure governing the conduct of oral hearings, the following general notes are offered for guidance. These notes are necessarily subject to revision in the light of experience and persons contemplating requesting an oral hearing are advised to seek the advice of the Secretaire at an early stage.

Guidance notes from CPMP on oral hearings

a) The competent authorities of the Member State which initially authorized the product will appoint someone to act as the rapporteur, and the applicant should liaise with them regarding:

- the time necessary for preparation of the written response to objections;
- date(s) for the likely consideration of the application by the CPMP to formulate the opinion;
- whether a hearing, or only a written response to objections may be more appropriate;
- the crucial issues from the objections that will need to be addressed in writing and/or at a hearing.

Five working days before the CPMP meeting, the rapporteur will be advised by the other concerned Member States whether their objections are satisfied by the written material. If any points remain, the applicant can then be advised and will be able to concentrate only on these issues.

b) It is important that persons preparing for and attending hearings bear in mind that the oral proceedings of the Committee are multi-lingual and that simultaneous technical interpretation during the hearing will be necessary. For this reason arguments of a very technical or scientific nature are better expressed in writing.

c) Any new written documents to be used in conjunction with a hearing should be distributed to the Members of the CPMP 30 working days before the meeting. In addition the Secretariat will require an additional 10 copies of any document to which detailed reference is to be made for the interpreters. Applicants should liaise with the original Member State regarding the content of written documents which are to be used in conjunction with a hearing.

d) Without wishing to specify a formal time limit; the CPMP considers that hearings lasting more than half an hour will not usually be necessary. Only in exceptional circumstances will the applicant be allowed to make a presentation to the CPMP (the duration will be at the discretion of the CPMP and its chairman). The hearing is held to allow clarification of outstanding issues by questioning from members of the CPMP. Depending on the issues raised in the reasoned objections of the Member States, it would normally be appropriate for between one and four persons to appear on behalf of the company concerned.

e) Persons attending hearings should notify the Secretariat in good time of the language in which they propose to express themselves so that arrangements for interpretation can be made.

f) Persons attending hearings are advised to concentrate their discussions on the remaining, unresolved objections rather than spending time on arguments which are not relevant to the reasoned objections of the Member States to an application for MA or to the grounds advanced by Member States for refusal, suspension or withdrawal of a MA.

g) Thereafter, the representatives will be asked to withdraw while the Committee discuses its opinion, which will be sent in writing to the company.

10.6.7 CPMP multi-state application sequence

Directive 83/570/EEC

1) A firm applying to use the multi-state procedure

   - consults the competent authority (rapporteur) which granted the initial authorisation, agreeing any additions to be made;
   - submits a complete dossier to the other Member States, concerned with the application, asking them to take into due consideration the initial authorisation (minimum of 2 other Member States);
   - notifies CPMP secretariat (Committee for Proprietary Medicinal Products).

2) CPMP secretaries sends a telex to all Member States, stating the name of the company and the product, the original country of authorisation and lists countries concerned. Concerned States are thereby invited to notify receipt of the dossier to the CPMP secretariat.

   All available assessment reports relating to the same product are immediately communicated by the competent authorities to the Member States concerned and to the Committee.

3) Concerned Member States confirm receipt of complete application (by telex or telefax), usually to the secretariat. When all concerned countries have responded, the 120 day period of consideration commences. Only one telex per month will be sent, notifying the commencement of the 120 day period for all multi-state applications in that month.

   An opinion of the CPMP is not required if no Member State puts forward any reasoned objection during the 120 day period.

4) In exceptional cases, where there are reasoned objections within the 120 days, the Member State concerned notifies them directly to the applicant and the original authorizing country. A copy is also sent to the CPMP secretariat for information.

5) After receiving all reasoned objections, the applicant enters into consultation with the rapporteur. The time needed by the applicant to respond is agreed. If an extension of the time is necessary, the applicant, in agreement with the rapporteur, informs the Secretariat of the Committee. The Secretariat thereafter informs the Member States.

6) In response to the reasoned objections, the applicant prepares a single response to each of the objections raised (questions and answers). The format of this response is to be the same sequence as the dossier.

7) The applicant liaises with the rapporteur regarding the scheduling of the CPMP hearing/opinion. Following agreement, the company's response is circulated to all members of the CPMP, by name, at least 30 working days before the meeting, in the format of Notice to Applicants.

8) If the company wishes an oral presentation (hearing), this should be confirmed to the Secretariat 30 working days in advance.

9) The rapporteur keeps the CPMP informed of the progress of the application.

10) At the CPMP meeting, the rapporteur reports on the resolution of objections. The hearing (if any) takes place. After discussion and/or hearing, the Secretariat drafts the Opinion, which is adopted by the Committee.

11) Within 60 days of the issue of the Opinion, the Member States concerned notify the Commission of their decision taken regarding the application.

10.7 The Special Community Concertation Procedure for the Marketing of High Technology Medicinal Products, Particularly those Derived from Biotechnology (Directive 87/22/EEC)

10.7.1 Purpose and scope of the special procedure

The rules governing the special Community procedure are set out in Council Directive 87/22/EEC of 22 December 1986 "on the approximation of national measures relating to the placing on the market of high technology medicinal products, particularly those derived from biotechnology" (O.J. L 15, of 12.1.1987, p. 38).

objective of the Directive

This Directive entered into force on 1 July 1987. The objective of the special procedure is to enable questions of principle relating to the quality, safety and efficacy of Medicinal Products developed by means of new biotechnology processes and other high technology medicinal products to be resolved at Community level within the CPMP before any national decision is reached concerning the marketing of the product concerned. Use of the procedure will, therefore, facilitate subsequent access to the markets of the other Member States. Moreover, the Commission will publish a list of the products in respect of which the procedure has been used and these will automatically benefit from the ten year period of protection of innovation afforded by Article 4 point 8 of Directive 65/65/EEC as amended by Council Directive 87/21/EEC of 22 December 1986 (O.JL 15 of 17.1.1987, p. 36).

The timetable for the concertation is set by the filing of the application in the first Member State, and it is their acceptance of it as a valid application which initiates them in their role as the rapporteur. Applications in other Member States should be made as soon as possible thereafter (normally not more than 2-3 weeks after the first Member State) to ensure that the assessments of the application can take place in parallel in all Member States concerned and that all problems can be fully discussed by the CPMP The concertation timetable will be circulated to all Member States by the rapporteur and notified to the applicant. If applications are made in various Member States over a period longer than that recommended above, the benefit of the Community concertation may be reduced, since some Member States can not take full part in the discussions.

The new procedure is obligatory for all medicinal products developed by means of the following biotechnological processes:

- recombinant DNA technology;

- controlled expression of genes coding for biologically active proteins in prokaryotes and eukaryotes, including transformed mammalian cells;

- hybridoma and monoclonal antibody methods.

It should be noted that the obligation to use the new Community procedure applies to all medicinal products which are developed by means of these biotechnological processes, irrespective of whether or not they are covered by existing Community Directives. Thus, for example, an application to market a new recombinant DNA vaccine in a Member State will have to be referred to the CPMP.

In accordance with Article 2 (3) of Directive 87/22/EEC, there is, however, no obligation to refer applications to market these biotechnology medicinal products to the CPMP if the applicant certifies that application for marketing authorisation is being made to one Member State only, that no other application for the product concerned has been made to the competent authorities of another Member State during the proceeding five years, and gives a firm undertaking not to seek authorisation in any other Member State for five years. Should an application for authorisation subsequently be made to another Member State, within the five year period, it will automatically be referred to the CPMP.

In addition, applicants for marketing authorisation for the following groups of products may request the competent authorities of the Member State concerned to refer the matter to the CPMP for consideration before any national decision on the application is reached; irrespective of whether or not they are currently covered by the Community Directives:

- medicinal products developed by other biotechnological processes which constitute a significant innovation;

- medicinal products administered by means of new delivery systems which constitute a significant innovation;

- medicinal products containing a new substance or an entirely new indication which is of significant therapeutic interest;

- new medicinal products based on radio-isotopes which are of significant therapeutic interest;

- medicinal products the manufacture of which employs processes which demonstrate a significant technical advance such as 2-dimensional electrophoresis under micro-gravity.

If the competent authorities are satisfied that the medicinal product has the sufficiently significant innovatory character which is claimed for it, the competent authorities shall refer the application to the CPMP. At the request of any Member State, the CPMP may consider the question of innovation and decide whether it is competent to consider the application. The Expert Reports accompanying such applications must therefore include a separate discussion and justification for the significant innovatory character claimed in respect of the product in comparison with other available therapies, processes or delivery systems. The time-limits for the examination of the application may start when the CPMP has agreed that the medicinal product is a significant innovation.

10.7.2 The procedure to be followed

procedure outlined in ;Notice to Applicants'

The concertation procedure, is designed to allow greater exchange of resource and discussion between Member States. Therefore the procedure is intrinsically flexible. The role and responsibility of the rapporteur Member State includes determination of the time-periods, as well as close liaison with the applicant and other concerned Member States. The procedure as outlined in a 'Notice to Applicants' is intended as a guide to applicants. Of course, with experience, the procedure may be improved.

The applicant should firstly liaise with the rapporteur Member State, who will refer the application to the CPMP in accordance with the special concertation procedure.

The applicant should immediately (within 2-3 weeks) submit a formal application, with a full dossier, to those other Member States in which an MA is being sought. To facilitate application, all Member States have agreed that the receipt of a full dossier in English is satisfactory for the purposes of starting the procedure, provided translations of appropriate parts are supplied to the competent authorities within 30 days.

submission of the dossier to CPMP

A complete dossier of the application must be transmitted to the Secretariat of the CPMP by the applicant and the Members of the Committee are permitted to consult this dossier at the Secretariat. It is suggested, particularly in the case of large dossiers, that contact be made with the Secretariat in advance, to arrange for receipt of the documents.

A summary of the dossier, consisting of at least the administrative information, the summary of the product characteristics and the Expert Reports and tabulated formats or equivalent documents in the case of products not yet covered by the Community Directives, must be sent to those Member States not directly concerned by the application. The applicant is recommended to involve all Member States and to make a copy of the full dossier available, upon request, to such competent authorities.

All information submitted to the Committee is strictly confidential.

The competent authorities of the Member State which referred the matter to the Committee act as rapporteur and ensure that the Committee receives all information relevant to the evaluation of the product.

The rapporteur Member State initiates the procedure as soon as all the Member States concerned have received the file.

time-scale for examination of applications

Examination of the application by the CPMP takes place at the same time as examination of the application by the authorities of the rapporteur Member State which initiated the procedure. In the case of concertation applications, no previous assessment by a competent authority will have been prepared. Therefore, bio/high technology applications are likely to be treated as exceptional and the period of review would therefore be 120 plus 90 days. These time limits may be suspended where the applicant is required to provide additional information, or to provide oral or written explanation. For its part, the CPMP is required to complete its consideration 30 days before the expiry of these time limits so that the national authorities have time to consider the opinion of the Committee before reaching a final decision. To this end, the Committee and the applicant are kept informed of any decision by the rapporteur to extend the time limit laid down or to suspend the running of time by requesting additional information from the firm.

Any additional documentation should be circulated to all members of the CPMP by name and/or to the relevant working parties, at least 30 working days before the meeting, in the same format as the dossier, otherwise consideration of the application may be postponed. Normally the applicant should send the consolidated response to all objections raised, within a fixed time-period - in principle within 3 months.

If the company wishes a hearing, this should be confirmed to the rapporteur Member State and the Secretariat 30 working days in advance by the company. Applicants should liaise with the rapporteur regarding the content of written documents which are to be used in conjunction with a hearing.

In addition to providing whatever written information it considers appropriate, the applicant may present oral submissions in person before the Committee. The applicant may request that the CPMP postpones consideration of the application in order to allow time for the preparation of written or oral submissions.

The opinion of the Committee will be concerned exclusively with the quality, safety and efficacy of the product concerned. The opinion, which is not binding, will be transmitted to the applicant and to the Member States. The Member States concerned by the application are required to reach a decision on the action they intend to take following the Committee's opinion within thirty days of its receipt. The other Member States will consider the opinion of the Committee when examining any subsequent application for authorisation in respect of the same product.

10.7.3 Submission of applications

In the first instance, the applicant should consult the competent authority of the rapporteur State. Only after this consultation should a complete application for marketing authorisation be submitted to the competent authorities of the other Member States concerned. Information on the number of copies of the application required and the language in which the information is to be presented is given in Table 10.5.

Where reference to the CPMP is compulsory, the application will automatically be referred by the competent authority unless the application is accompanied by a signed declaration that no other application for the product concerned has been made to the competent authorities of another Member State during the preceding 5 years, and that no application will be made to the competent authorises of another Member State during the next 5 years in respect of the product concerned. Applicants are therefore advised to send the summaries of the dossier directly to the competent authorities of the Member States who are not directly concerned at the same time as they submit the full dossier to the rapporteur Member State. A complete copy of the dossier should also be sent to the CPMP Secretariat at this time.

Where a firm wishes to submit an application for a MA for a high technology medicinal product for consideration by the CPMP on a voluntary basis , it should include with the application for MA a request that the competent authority concerned submit the application to the CPMP pursuant to Article 2 of Directive 87/22/EEC. As soon as the agreement of the competent authority has been obtained, the firm should send the relevant documentation directly to the competent authorities of the Member States and to the Secretariat of the Committee.

10.7.4 Presentation of applications

So far as possible, applicants should follow the guidance on the presentation of dossiers and the preparation of Expert Reports provided by CPMP. However it is recognised that this guidance may not be entirely appropriate for applications concerning vaccines, toxins, sera and radioactive isotopes which are currently excluded from the scope of the Community pharmaceutical directives by Article 34 of Directive 75/319/EEC. Persons preparing applications for authorisation for such products are therefore advised to discuss the presentation of their dossier with the competent national authority at an early stage.

10.7.5 CPMP concertation application sequence - Directive 87/22/EEC

1) A firm applying to use the concertation procedure;

- for Biotechnology: requests the first Member State to act as rapporteur;
- for other High Technology: also requests the first Member State to accept the application as suitable for the procedure (the Member State may refer the matter to the CPMP for agreement).

2) The company makes a formal application to the first Member State who acts as rapporteur thereafter, informs the first Member State about the other Member States concerned and notifies the CPMP of the application. The rapporteur sends a telex to all other Member States, informing them of the application.

3) The company makes a complete application to as many other Member States as possible and supplies (at least) a summary of the dossier to all other States, certifying that dossiers/summaries are identical. A full dossier plus a summary is supplied to the CPMP.

4) The commencement of the procedure is determined by the rapporteur Member State. Prior to commencement, all Member States should have received the dossier/summary. Member States encountering difficulty regarding receipt of a dossier/summary notify the rapporteur directly. Up to 10 working days may elapse between receipt of the dossier and commencement of the procedure, in order to validate the dossier. To facilitate applications, all Member States have agreed that the receipt of a full dossier in English is satisfactory for the purposes of starting the procedure, provided translations of appropriate parts are supplied to the competent authorities within 30 days.

5) The rapporteur establishes a time-table for review as in the Directives (for Bio/High Technology Applications which may be considered as exceptional, the period of review would usually be 120 plus 90 days). The time-table is communicated by the rapporteur to all members and the Secretariat of the CPMP and to the company. The Secretariat informs the appropriate working parties.

6) The rapporteur Member State prepares an evaluation report with questions, circulates it to all Member States, to the Secretariat of the CPMP and the questions are sent to the company within 45 days of the commencement of the procedure, and stops the "clock". By agreement, within a further 60 days, all other Member States are invited to add comments/questions to this.

7) The rapporteur Member State makes a compilation of all of the objections and these are discussed/filtered by the appropriate working party and CPMP. By agreement, within a further 45 days, the resulting list of objections is passed by the rapporteur to the company, and is circulated to all Member States.

8) The company sends to all Member States a single answer to all the questions, before the expiry of a fixed time period, agreed with the rapporteur - in principle within 3 months. The rapporteur starts the clock again and will confirm to the company whether, by virtue of the exceptional nature of the application, an additional 90 days will be added to the period of review. The company liaises with the rapporteur State regarding discussions/hearings with the CPMP or its working parties.

9) 30 days after the restart of the clock, all the Member States concerned send to the rapporteur and the CPMP their conclusions on the answers to the questions raised. An appointment for the date of the opinion by the CPMP, which must be 30 days before the expiry of the overall time-limit, is made by the rapporteur.

10) Any additional documentation in relation to the opinion should be circulated to all members of the CPMP, by name, at least 30 working days before the meeting, in the format of Notice to Applicants.

11) At the CPMP meeting, the rapporteur State reports on the resolution of objections. The hearing (if any) takes place. The secretariat drafts the opinion after the discussion and/or hearing which the Committee adopts the second day.

12) Within 30 days of the issue of the opinion, the rapporteur State and other Member States concerned notify the Commission of their decision on the action to be taken following the opinion of the Committee.

13) The member states not directly concerned inform the Committee of subsequent applications for marketing authorisation.

14) Member States inform the Committee in advance of any new regulatory action on phamacovigilance matters or in urgent cases, immediately thereafter.

10.8 Recommended additional reading

Cook, T., Doyle, C & Jabbari, D. (1991). Pharmaceuticals, Biotechnology and the Law. MacMillan, Publishers: Basingstoke, UK. ISBN 0.33-5116-9.

Manufacture and evaluation of medicinal products produced by biotechnological processes

Appendix 11.1

Manufacture and evaluation of medicinal products produced by biotechnological processes

11.1 Introduction

In the previous chapter, we provided information concerning the procedures that are to be followed in making an application for market authorization. We indicated that a substantive dossier needs to be submitted to "competent authorities". Much of this dossier deals with ensuring that the properties of the product have been properly determined and that the manufacturing practices used ensure the quality of the product.

Here we will examine the issues of Good Laboratory Practice (GLP), Good Manufacturing Practice (GMP), Quality Control (QC) and Quality Assurance (QA).

11.2 Good laboratory practice (GLP)

As we saw in chapter 4 Good Laboratory Practice (GLP) guides give a general description of the practice, procedures and conditions that should be maintained within laboratories used for evaluating the hazards to health of chemical substances and the development and testing of medicines. The GLP guides cover such items as laboratory design, equipment and facilities, management structure, operating procedures, qualified persons, education and training of staff, record keeping and quality assurance and independent auditing of the activities of the laboratory and the data producer.

purposes of GLP

The main purposes of Good Laboratory Practice is to promote the development of quality data generation that is mutually acceptable to a wide range of countries and institutions. The application of agreed laboratory procedures should avoid the need for duplicative testing and reduce technical barriers to trade. They should also improve the protection of human health and the environment.

GLP is concerned with all laboratory practice including the organisational process and the conditions under which laboratory studies are planned, performed, monitored, recorded and reported. The GLP guides define, for instance the responsibilities of individuals from the Director downwards and they provide conditions that the laboratory facility should fulfil including the need to establish regular inspection of apparatus and equipment, its maintenance and calibration. All routine procedures and practices must be enshrined in standard operating procedures which effectively set the standards for the laboratory. Particular emphasis is given to the importance of generating quality data and maintaining its integrity during processing (see Chapter 4 for further details).

11.3 Good manufacturing practice (GMP)

Good Manufacturing Practice (GMP) guides are similar to GLP guides except they relate to the process of manufacturing. The holder of a Manufacturing Authorization must manufacture medicinal products so as to ensure that they are fit for their intended use and comply with the requirements of the Marketing Authorization. GMP is concerned with both production and quality control (QC).

The basic requirements of GMP are:

- manufacturing processes are clearly defined, systematically reviewed and shown to be capable of consistently manufacturing medicinal products of the required quality and complying with their specification;

- critical steps of manufacturing processes and significant changes to the process are validated;

- all necessary facilities for GMP are provided (including qualified personnel, adequate space, suitable equipment and services, correct materials, containers and labels, suitable storage and transport);

- instructions and procedures are written in a clear, specific and unambiguous language;

- operators are trained to carry out procedures correctly;

- records are made (manually or by instrumentation) during manufacture which demonstrate that all the steps required by the defined procedures and instruction were taken and that the quantity and quality of the product was as expected. Any significant deviations are fully recorded and investigated;

- records must be kept of manufacture and distribution which enable the complete history of a batch to be traced. Such records must be retained in a comprehensible and accessible form;

- the distribution of the product minimises any risk to its quality;

- a system must be developed to recall any batch of product;

- complaints about products must be examined. Any quality defects must be investigated and appropriate measures taken to prevent re-occurrence.

Having read this list carefully you will notice that special emphasis is again placed on maintaining the quality of the product and on keeping records of the manufacturing process and the distribution of the product. GMP also includes some items which ensure the quality of the product after it has left the factory.

As a reference source, we have produced a summary the EEC Guide to Good Manufacturing Practice (issued January 1989) as an appendix to this chapter.

11.4 Quality control (QC)

The requirements for Quality Control QC are:

- adequate facilities, trained personnel and approved procedures are available for sampling, inspecting and testing starting materials, packaging materials, intermediate, - bulk - and finished products. Where appropriate these must be adequate to monitor environmental conditions for GMP purposes;

- samples for QC analysis are taken in an approved manner;

- test methods are validated;

- records are made which demonstrate that all required sampling, inspection and test procedures were actually carried out. Any deviations to be recorded and investigated;

- the finished product must contain active ingredients complying with the qualitative and quantitative composition of the Marketing Authorisation. It must be of the required purity, enclosed in its proper container and correctly labelled;

- records are made of the results of inspection and testing and that these are formally assessed against specification;

- no batch of product is released for sale or supply prior to certification by a Qualified Person that is not in accordance with the requirements of the Marketing Authorisation;

- sufficient reference samples of the starting materials and products are retained to permit future examination of the product.

GMP also provides details of what should be included in the specifications used in the manufacture of a product. The reader should be alerted to the fact that specifications are included in the authorisation to market a product. Specifications have to be established for starting materials, packaging, intermediate, bulk and finished products.

11.5 Quality Assurance (QA)

Many people are confused by the terms Quality Assurance and Quality Control, so let us see if we can make the distinction clear. Quality Assurance (QA) is a wide ranging concept which covers all matters which individually and collectively influence the quality of a product. It is the total of all the arrangements made to ensure that a product is of the quality for its intended use. QA therefore incorporates GLP, GMP and QC.

We can therefore represent the relationship of QA, GLP, GMP and QC as shown in Figure 11.1.

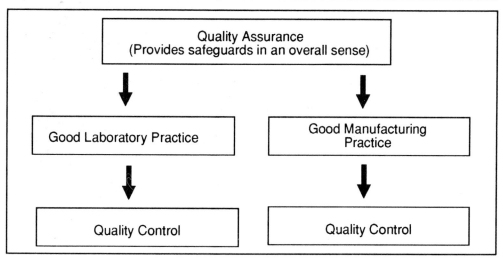

Figure 11.1 The relationship between QA, GLP, GMP and QC.

11.6 Market authorisation

In Chapter 10, we indicated that a large dossier of information needs to be submitted to the licensing authorities before authorisation to market a product could be gained. We learnt in subsequent sections some of the details of Directives and Guidelines that need to be followed in the development and production processes. In this, chapter we examine the contents of the dossier that needs to be submitted to the licensing authorities.

Table 11.1 provides a summary of the kind of information which is included in a licence dossier. You will notice some of these items have already been discussed under the headings of GLP, GMP and QC. We do not wish to dwell on them but to think about the rationale behind the production of the dossier.

Contents	Remarks
Premises (i.e buildings)	These relate to GMP and GLP and are mainly monitored by inspection
Personnel	
organisation	
education and training	
Process validation	
reduction of contaminants	
Reproducibility of quality	
Source materials	These relate to GMP and are subject to QC.
chemicals	
cultures	
Production procedures	The evaluation of these are part of GMP and are subject to QC
bulk product	
final product	
filing lot	
Quality control	Subject to GLP
safety	
potency	
reprodcibility	
Stablity	Subject to QC
Reprocessing	Subject to QC
Pre-clinical testing	
Clinical testing	

Table 11.1 The information contained in a licence application dossier.

The main objectives of the dossier are to provide the licensing authority with documentary evidence that the product fulfils the requirements of all of the relevant regulations concerning the efficacy and safety and that the procedures adopted for the production and evaluation of the product provide an assurance of quality. The onus therefore is placed on the manufacturer to produce a specification for the product and process which fulfils these requirements.

Let us examine each of the headings provided in Table 11.1 in turn to learn more of what is involved.

11.6.1 Premises

clean

We have learnt in our discussion of GMP that premises must be designed and constructed to facilitate effective cleaning and maintenance. Many biotechnological medicinal products cannot be autoclaved and so have to be manufactured under aseptic conditions, involving classified rooms with air of appropriate quality. Some care has to be taken that the product is not a vehicle for transmitting infection.

11.6.2 Personnel

education
training

A scheme defining the responsibilities for personnel in charge of production, quality control and quality assurance should be included in the licensing file. The same individual should not be responsible for both production and QC to avoid possible conflicts of interest.

There must be sufficient qualified personnel to carry out all the tasks involved in production. This implies a commitment to education and training on the part of manufacturers.

11.6.3 Process validation

Reduction of contaminants

A basic principle of GMP is that the maintenance of quality requires not only extensive testing of the final product but also high standards throughout the process of manufacture. This may be illustrated by reference to the problem of removal of bacteria by filtration. It can be very difficult to detect low levels of bacteriological contamination, as illustrated in the following table (Table 11.2) and so it is essential to reduce to a minimum any possibility of contamination prior to the final filtration.

The data of Table 11.2 is based on testing 20 vials in each batch.

Number of contaminated vials	Batch Size		
	1000	10 000	100 000
1	2%	0.2%	0.02%
10	20%	2.2%	0.22%
100	88%	18.4%	2.0%

Table 11.2 Chance of detecting contamination in one vial if twenty vials are tested from each batch.

The data show that if only 1 vial is contaminated in every batch of 1000 vials and 20 vials are tested, then there is only a 2% chance of detecting the contamination. If the batch size is 100 000 vials, then the chances of detecting a single contaminated vial if 20 vials are tested is only 0.02%.

spiking
experiments

Eliminating pathogenic viruses, a potential risk when mammalian cells are used as a source of material, is even more difficult. Because of the relative insensitivity of many assays for viruses, the reduction factor in each purification step should be established for a number of viruses. This is done by trial experiments in which known quantities of virus are added before carrying out the purification scheme (ie 'spiking' experiments). When using source materials proven to be free of contaminants (like viruses), reduction factors should be in the range 10^5 to 10^{10}. The objective of the validation of the reduction is a precaution against as yet 'unknown' viruses that could conceivably be present.

Reproducibility of quality

five consecutive batches

Chemical by-products arising from the growth of cells cover a wider range than those accompanying chemical synthesis. Sometimes they can only be detected by bioassays, and so establishing reproducibility of quality for a GMP can be quite difficult. Commonly, five consecutive batches are produced to demonstrate consistency before submitting a licensing.

Here we have included only three types of process validation studies, two relating to removal of contaminants, one relating to reproducibility of quality. These however illustrate the principles involved.

The reader should recognise that the performance of such studies enables the manufacturer to make certain specifications for the product. For example, it could be specified that the product would contain less than a certain number of viruses or that fewer than a specified proportion of the product would contain bacteria.

Source materials

It is obvious that source materials should be of high quality, using, when appropriate, guidance given by pharmacopoeias.

seed lot system

The biotechnological production of medicines using, cells containing and expressing deliberately inserted or amplified genetic material requires special consideration. We must prove the cultures' purity. Cultures are grown to a suitable density, dispensed into containers that allow a long period of storage (usually in liquid nitrogen) without jeopardising the quality. Subsequently, the cells are tested for contaminants such as bacteria, viruses and mycoplasma. Each time a new batch of product is required, one or more containers of the stock culture is used to seed the fermenter (hence 'seed lot'), ensuring as far as is possible consistency in the source of the biological material. This practice is widely accepted in the production of vaccines, monoclonal antibodies and recombinant DNA products.

11.6.4 Production procedures

batch and continuous culture

Products can be made one batch at a time or by continuous culture. Either way, the description of the method of cultivation should demonstrate clearly that contamination with other products made in the same premises cannot occur. Again this can be built into the specification of the product.

In batch production, cells are cultivated until reasons related to economy or formation of contaminants necessitate stopping the process and then the cells are harvested. During continuous production, the cells remain in the fermenter while nutrients flow into and product out of the fermenter. A higher yield may be possible, but it is likely to be more difficult to prove that the product is consistent in quality.

Before we turn our attention to the tests that may be conducted during the production process, let us first make clear the main stages of product processing, after the cultivation stage. We can divide this into three main phases. The production of the bulk product, the final product and the filing lot. We define these as follows:

bulk product

Bulk product. After harvesting, the cells are ruptured if the product is intracellular and cell debris is discarded. The preparation at this stage is referred to as the bulk product. It is appropriate to test for unwanted materials (eg viruses) for which relatively insensitive assay methods are available. If 'absent' from the bulk product, there is a further safety margin provided by subsequent purification and/or inactivation procedures.

final bulk   **Final bulk.** After purification and possible inactivation, the final - or purified - bulk product is obtained. It is at this stage that most of the tests such as potency, chemical and biological characterisation etc. are performed. We will briefly deal with a few of the more common ones a little later.

filling lot   **Filling lot.** After dispensing into the final containers, the product is referred to as the final lot or filling lot. At this stage, tests for sterility, safety and potency are carried out. It is not our intention here to examine all of the tests that may be carried out on the various stages described above. These tests will be predominately concerned with the physical and chemical identity and with the purity of the product at each stage. Thus for each stage, the quality of the product must be specified. If, at any time this specification is not achieved, the product must be rejected or reprocessed. The tests will also involve evaluating the potency of the product and determining the level of contamination. With products generated by biotechnology, particular concern is attached to the presence of residual DNA and viruses in the product. Thus, we will briefly consider the potency, DNA and virus tests that may be conducted on the product.

Potency tests

Wherever possible, the potency of a product should be tested against an international reference preparation. The goal should be a product that differs from the standard only in terms of concentration, although this might require qualification depending on the stability of the standard during storage.

There are four categories of potency tests: animal, tissue culture, immunological and chemical.

challenge potency tests   Animal tests, especially challenge potency tests, simulate the wanted effect relatively well. In the case of many vaccines, animals are immunised with dilutions of either the product or the reference preparation. The animals are subsequently challenged with the agent against which the vaccine should provide protection. The surviving animals are counted and the potency is calculated using a probit distribution.

tissue cultures and challenge potency test   Tissue cultures can be used as an alternative to the challenge potency test. Animals are immunised with the vaccine and the resulting antibodies used to neutralise a fixed amount of the toxin. The mixture of antibodies and toxin is added to a tissue culture to establish the remaining cytotoxic effect. These tests use fewer animals, are easier to standardise and cause less harm to animals. Likewise, tissue culture tests can be used to establish the potency of live viral vaccines. The tissue culture is incubated with dilutions of the vaccine; the number of plaques of cells infected with the virus establishes the potency of the preparation.

immunological tests   Immunological tests can involve immuno-electrophoresis, immuno-diffusion or ELISA (Enzyme Linked Immune Sorbant Assays). In immuno-electrophoretic and immuno-diffusion tests, the antiserum is added to the gel and during electrophoresis/diffusion of the product and the reference standard, precipitations occur at equimolar concentrations. The size of the peaks (in rocket electrophoresis) or rings is proportional to the concentration of the product.

Chemical potency tests make extensive use of HPLC, (High Performance Liquid Chromatography) and other biochemical separation techniques.

DNA tests

The presence of DNA is of particular concern in many biotechnology products. The amount of residual DNA can be established by a nucleic acid hybridisation assay, the detection limit for a repetitive sequence being of the order of 10^{-12}g (1 pg). This is a sensitive indicator of the presence of a host cell DNA in a product derived from those cells. However there is still concern whether it provides adequate 'protection' against putative oncogenic DNA sequences. Tests with oncogenic viral DNA in animals suggest that the risk of a tumorigenic event after receiving 100 pg of DNA is of the order of 1 in 5-20 million. There is considerable uncertainty in the assumptions made in the calculations. In general, it is advised that the DNA present per dose should be no more than 10-100 pg.

Virus tests

Several techniques are available for establishing that products are not contaminated with viruses. Commonly used are:

- inoculation, as appropriate to the virus, of chick embryos, suckling or adult mice or higher species;

- inoculation as above and testing for the production of antibodies against specific viruses with ELISA or immuno-fluorescence assays. Amounts of viruses just sufficient to cause propagation of the virus can be detected;

- cell cultures can be used to test for viruses that are cytopathogenic.

11.6.5 Quality control of the final product

A medicinal product should be free of toxic side-effects (i.e. be safe), effective (i.e. be potent) and be prepared consistently. The conditions of GMP, quality control and clinical testing should all support the above goals. Each filling lot, (occasionally the final bulk), should be evaluated by the following test.

Safety

In a general test for safety, a relatively high dose is inoculated intraperitoneally into, for example young guinea pigs and young mice. After one or two weeks, the animals should not show any ill effects such as loss of weight relative to control animals.

LAL test

Pyrogens, usually bacterial fragments, are detected by injecting three rabbits intravenously and recording any rise in temperature. An alternative assay uses an extract of amoebocytes from the horseshoe crab (*Limulus polyphemus*; called the LAL test. The extract forms a gel-clot with pyrogens. Not surprisingly, the rabbit and LAL tests for pyrogens do not always coincide and require careful validation.

Reproducibility

trend analysis

The results of potency and other tests on consecutive batches should be submitted to trend analysis. Any undue trend should be investigated and correcting action taken in accordance with contingency plans.

microbial contamination

Other safety tests include screening for bacteria, viruses, mycoplasma and DNA. (These tests greatly overlap with the tests conducted during the production process).

Stability

expiry term

Samples of several (preferably all) batches should be stored at the appropriate temperatures during the expiry term (the maximum time during which the product can be stored without serious loss of quality). The potency should not be lower than the claimed potency at the time of issue. Often additives are included in the final formulation of a product to improve stability. The reader should note however that these additives and the potency are both included in the specification of the product.

If there is a possibility that toxic products can be formed during storage, it must be shown that the concentrations of these products do not exceed safety limits by the full expiry time.

11.6.6 Reprocessing

main stream
side stream

Reprocessing is a rescue operation used once a product (or intermediate stage in its production is out of specification. It may involve repeating a production step or applying an additional procedure in order to regain the specification. Reprocessing should only be allowed if it has been validated and described in the licensing file. In 'main stream' reprocessing a procedure is repeated on fractions of an intermediate product, whilst 'side stream' reprocessing involves handling fractions that would normally be discarded. The latter carries more risk of changing the specification of the product.

Side stream reprocessing often involves purification processes which are different from those conducted in the main manufacturing stream. These different processes could therefore make some alterations in the final product.'

11.7 Pre-clinical and Clinical Testing

The testing of medical substances is done according to a variety of codes of practice. Most of these are produced by individual states and provide detailed, authoritative and practical accounts. The European Commission proposals as to Good Clinical Practice are also provide recommendations as to the design and protocols of studies. In this section, we provide a brief overview of the clinical testing of medicines since this is of general application rather than being specifically designed for the products of biotechnology.

The testing of medicines can be divided into five phases.

These are:

- pre-clinical testing;

- phase I tests;

- phase II tests;

- phase III tests;

- phase IV tests;

We will consider the purposes of each phase

11.7.1 Pre-clinical testing

This phase aims to determine whether the substance is biologically active and begins to form a view as to its safety. Such testing involves extensive testing using animals. It is therefore subject to the conditions described in Chapter 7.

Referring specifically to biotechnology products, five groups of compounds can be especially distinguished:

- hormones;

- cytokines and other regulatory factors;

- blood products;

- monoclonal antibodies;

- vaccines.

Clearly such products may exhibit high pharmacological activity or influence the immune or other physiological system. Great care, therefore is needed in carrying out pre-clinical testing of such materials. Such work typically takes one or two years. It is usual, however, for trials on animals to continue in order to look for long-term effects.

11.7.2 Clinical testing - Phase 1 Tests

The key EC-Directives are: 75/318/EEC on the approximation of the laws of Member States relating to analytical, pharmacotoxicological and clinical standards and protocols in respect of the testing of proprietary medicinal products - amendments to 75/318/EEC, were introduced by Directives 83/570/EEC and 87/19/EEC.

Phase I tests begin the process of testing using humans. The objectives of this phase are the development of the pharmacological profile of the product and the initiation of safety studies in humans.

This phase determines the pharmacological activity of the product, its safe dosage and ascertains whether, and to what extent, the product may have harmful effects.

These tests involve a small number of normal, healthy volunteers (that is non-patients). The tests aim to provide information concerning dose rates, blood levels, metabolism of the drug and toxicology.

It should be noted that generally no statutory control applies to this stage. This is consistent with EC-Directive 65/65/EEC (amended by Directive 89/341/EEC, appendices 11 and 18). These make it clear that the control exercised over medicinal products does not apply to medicinal compounds intended for research or development trials. National positions on this are generally consistent with this EC overview. In the UK for examples, the relevant Act (Medicines Act 1968) states that:

'...in this act' medicinal product' does not include any substance or article which is manufactured for use wholly or mainly by being administered to one or more human beings or animals, where it is to be administered to them'

a)   *in the course of the business of the person who has manufactured it (in this subsection referred to as 'the manufacturer'), or on behalf of the manufacturer in the course of the business of a laboratory or research establishment carried on by another person, and*

b)   *solely by way of a test for ascertaining what effects it has when administered, and*

c)   *in circumstances where the manufacturer has no knowledge of any evidence that those effects are likely to be beneficial to those human beings, or beneficial to, or otherwise advantageous in relation to, those animals, as the case may be,*

And which (having been so manufactured) is not sold, supplied or exported for use wholly or mainly in any way not fulfilling all the conditions specified in paragraphs (a) to (c) of this subsection.'

11.7.3 Clinical testing - Phase II tests

These are efficacy studies usually involving 200-300 volunteer patients. Note that these are the first studies to involve patients and usually take place over about two years. Animal and healthy human studies continue during this phase. Since these studies involve patients, these studies come within the requirements of EC-Directives 65/65/EEC, 89/341/EEC.

11.7.4 Clinical testing - Phase III tests

This phase involves extensive clinical trials involving usually about 2000 volunteer patients. These trials provide a comprehensive determination of efficacy and harmful effects (side effects) and provides further information concerning dose rates. Usually these trials compare the product with a placebo (a formulation not having an active ingredient) and are usually carried out in a normal clinical environment. Often 'double blind' in which the identity of the product and placebo samples is hidden from those conducting the trial and from the volunteer patients.

11.7.5 Clinical testing - Phase IV tests

This is a rather poorly defined term which relates to clinical trials which are extensions of phase III trials. These are conducted after the product market authorization (registration) has been obtained. The concept of phase IV tests is not generally accepted and may cover a variety of different clinical trials. In some cases, they may be designed at obtaining further clinical information. In others, however, the aim may be more to promote the product.

11.7.6 The EC and the Regulatory control of Clinical Trials

Generally, the aim of EC-Directives is to harmonise the Procedures and Regulations of Member States. As yet there are no EC measures harmonising the controls on clinical testing. The influence of the EC is mainly on the development of the concepts of GLP, GMP, and QC, discussed elsewhere in this text. The emphasis of The EC has also been on harmonising the applications for market authorization. The relevant EC-Directives, however, specify that clinical trial data must be submitted and must conform to certain standards.

Appendix 11.1

A11.1 Summary of the EC Guide to good manufacturing practice

Here we provide a summary to the EC Guide to Good Manufacturing Practices. It is, of course, imperative that before a manufacturing process is contemplated that the full text of the Guide is examined and its guidance adhered to. You should anticipate that the Guide will be subject to revision and updating.

A11.2 Structure of the guide

The Guide to GMP is divided into several chapters dealing with:

- quality management (Chapter 1);

- personnel (Chapter 2);

- premises and equipment (Chapter 3);

- documentation (Chapter 4);

- production (Chapter 5);

- quality control (Chapter 6);

- contract manufacture and analysis (Chapter 7);

- complaints and product recall (Chapter 8);

- self-inspection (Chapter 9).

It also provides a range of supplementary guidelines for the manufacture of sterile medicinal products. The Guide to GMP has a glossary defining key terms used within the Guide. This glossary is useful so we will include it here.

A11.3 Glossary

Definitions given below apply to the words as used in this guide. They may have different meanings in other contexts.

Air Lock

An enclosed space with two or more doors, and which is interposed between two or more rooms, e.g. of differing class of cleanliness, for the purpose of controlling the air-flow between those rooms when they need to be entered. An air-lock is designed for the used by either people or goods.

Batch (or lot)

A defined quantity of starting material, packaging material or product processed in one process or series of processes so that it could be expected to be homogeneous.

Batch number (or lot number)

A distinctive combination of numbers and/or letter which specifically identifies a batch.

Bulk Product

Any product which has completed all processing stages up to, but not including, final packaging.

Calibration

The set of operations which establish, under specified conditions, the relationship between values indicated by a measuring instrument or measuring system, or values represented by a material measure, and the corresponding known values of a reference standard.

Clean Area

An area with defined environmental control of particulate and microbial contamination, constructed and used in such a way as to reduce the introduction, generation and retention of contaminants within the area. Note: The different degrees of environmental control are defined in the Supplementary Guidelines for the Manufacture of Sterile Medicinal Products.

Cross contamination

Contamination of a starting material or of a product with another material or product.

Finished product

A medicinal product which has undergone all stages of production, including packaging in its final container.

In-Process Control

Checks performed during production in order to monitor and if necessary to adjust the process to ensure that the product conforms its specification. The control of the environment or equipment may also be regarded as a part of in-process control.

Intermediate product

Partly processed material which must undergo further manufacturing steps before it becomes a bulk product.

Manufacture

All operations of purchase of materials and products, production, Quality Control, release, storage, distribution of medicinal products and the related controls.

Manufacturer

Holder of a Manufacturing Authorisation.

Medicinal Product

Any substance or combination of substances presented for treating or preventing disease in human beings or animals.

Any substance or combination of substances which may be administered to human beings or animals with a view to making a medical diagnosis or to restoring, correcting or modifying physiological functions in human beings or in animals is likewise considered a medicinal product.

Packaging

All operations, including filling and labelling, which a bulk product has to undergo in order to become a finished product.

Note: Sterile filling would not normally be regarded as part of packaging, the bulk product being the filled, but not finally packaged, primary containers.

Packaging material

Any material employed in the packaging of a medicinal product, excluding any outer packaging used for transportation or shipment. Packaging materials are referred to as primary or secondary according to whether or not they are intended to be in direct contact with the product.

Procedures

Description of the operations to be carried out, the precautions to be taken and measures to be applied directly or indirectly related to the manufacture of a medicinal product.

Production

All operations involved in the preparation of a medicinal product, from receipt of materials, through processing and packaging, to its completion as a finished product.

Qualification

Action of proving that any equipment works correctly and actually leads to the expected results. The word validation is sometimes widened to incorporate the concept of qualification.

Quarantine

The status of starting or packaging materials, intermediate, bulk or finished products isolated physically or by other effective means whilst awaiting a decision on their release or refusal.

Reconciliation

A comparison, making due allowance for normal variation, between the amount of product or materials theoretically and actually produced or used.

Recovery

The introduction of all or part of previous batches of the required quality into another batch at a defined stage of manufacture.

Reprocessing

The reworking of all or part of a batch of product of an unacceptable quality from a defined stage of production so that its quality may be rendered acceptable by one or more additional operations.

Return

Sending back to the manufacturer or distributor of a medicinal product which may or may not present a quality defect.

Starting Material

Any substance used in the production of a medicinal product, but excluding packaging materials.

Sterility

Sterility is the absence of living organisms. The conditions of the sterility test are given in the European Pharmacopoeia.

Validation

Action of proving, in accordance with the principles of Good Manufacturing Practice, that any procedure, process, equipment, material, activity or system actually leads to the expected results (see also qualification).

A11.4 Quality management (Chapter 1 of the Guide)

Principle

The holder of a Manufacturing Authorisation must manufacture medicinal products so as to ensure that they are fit for their intended use, comply with the requirements of the Marketing Authorisation and do not place patients at risk due to inadequate safety, quality or efficacy. The attainment of this quality objective is the responsibility of senior management and requires the participation and commitment by staff in many different departments and at all levels within the company, the company's suppliers and the distributors. To achieve the quality objective reliably there must be a comprehensively designed and correctly implemented system of Quality Assurance incorporating Good Manufacturing Practice and thus Quality Control. It should be fully documented and its effectiveness monitored. All parts of the Quality Assurance system should be adequately resourced with competent personnel, and suitable and sufficient premises, equipment and facilities. There are additional legal responsibilities for the holder of the Manufacturing Authorisation and for the Qualified Persons(s).

• The basic concepts of Quality Assurance, Good Manufacturing Practice and Quality Control are inter-related. (See sections 11.2, 11.3, 11.4 and 11.5 in the main body of text).

A11.5 Personnel (Chapter 2 of the Guide)

Principle

The establishment and maintenance of a satisfactory system of quality assurance and the correct manufacture of medicinal products relies upon people. For this reason there must be sufficient qualified personnel to carry out all the tasks which are the responsibility of the manufacturer. Individual responsibilities should be clearly understood by the individuals and recorded. All personnel should be aware of the principles of Good Manufacturing Practice that affect them and receive initial and continuing training, including hygiene instructions, relevant to their needs.

General

2.1   The manufacturer should have an adequate number of personnel with the necessary qualifications and practical experience. The responsibilities placed on any one individual should not be so extensive as to present any risk to quality.

2.2   The manufacturer must have an organisation chart. People in responsible situations should have specific duties recorded in written job descriptions and adequate authority to carry out their responsibilities. Their duties may be delegated to designated deputies of a satisfactory qualification level. There should be no gaps or unexplained overlaps in the responsibilities of those personnel concerned with application of Good Manufacturing Practice.

Key Personnel

2.3   Key Personnel includes the head of Production and the head of Quality Control.

The head of the Production Department generally has the following responsibilities:

 i)   to ensure that products are produced and stored according to the appropriate documentation in order to obtain the required quality;

 ii)   to approve the instructions relating to production operations and to ensure their strict implementation;

 iii)   to ensure that the production records are evaluated and signed by an authorised person before they are sent to the Quality Control Department;

 iv)   to check the maintenance of his department, premises and equipment;

 v)   to ensure that the appropriate validations are done;

 vi)   to ensure that the required initial and continuing training of his department personnel is carried out and adapted according to need.

The head of the Quality Control Department generally has the following responsibilities:

    i)   to approve or reject, as he sees fit, starting materials, packaging materials, and intermediate, bulk and finished products;

    ii)   to evaluate batch records;

    iii)   to ensure that all necessary testing is carried out;

    iv)   to approve specifications, sampling instructions, test methods and other Quality Control procedures;

    v)   to approve and monitor and contract analysts;

    vi)   to check the maintenance of his department, premises and equipment;

    vii)   to ensure that the appropriate validations are done;

    vii)   to ensure that the required initial and continuing training of his department personnel is carried out and adapted according to need.

Other duties of the Quality Control Department are summarized in Chapter 6.

The heads of Production and Quality Control generally have some shared, or jointly exercised, responsibilities relating to quality. These may include, with respect to national regulations:

- the authorisation of written procedures and other documents, including amendments;
- the monitoring and control of the manufacturing environment;
- plant hygiene;
- process validation;
- training;
- the approval and monitoring of suppliers of materials;
- the approval and monitoring of contract manufactures;
- the designation and monitoring of storage conditions for materials and products;
- the retention of records;
- the monitoring of compliance with the requirements of Good Manufacturing Practice;
- the inspection, investigation, and taking of samples, in order to monitor factors which may affect product quality.

Training

The manufacturer should provide training for all the personnel whose duties take them into production areas or into control laboratories (including the technical, maintenance and cleaning personnel), and for other personnel whose activities could affect the quality of the product. Besides the basic training on the theory and practice of Good Manufacturing Practice, newly recruited personnel should receive training appropriate to the duties assigned to them. Continuing training should also be given, and its practical effectiveness should be periodically assessed. Training programs should be available, approved by either the head of Production or the head of Quality Control, as appropriate. Training records should be kept.

Visitors or untrained personnel should, preferably, not be taken into the production and quality control areas. If this is unavoidable, they should be given information in advance, particularly about personal hygiene and the prescribed protective clothing. They should be closely supervised.

Personnel Hygiene

In summary, these measures include:

* detailed hygiene programmes should be established and adapted to the different needs within the factory;

* all personnel should receive medical examination upon recruitment. After the first medical examination, examinations should be carried out when necessary for the work and personal health;

* steps should be taken to ensure as far as is practicable that no person affected by an infectious disease or having open lesions on the exposed surface of the body is engaged in the manufacture of medicinal products;

* every person entering the manufacturing areas should wear protective garments appropriate to the operations to be carried out;

* eating, drinking, chewing or smoking, or the storage of food, drink, smoking materials or personal medication in the production and storage areas should be prohibited;

* direct contact should be avoided between the operator's hands and the exposed product as well as with any part of the equipment that comes into contact with the products;

* personnel should be instructed to use the hand-washing facilities;

A11.6 Premises and equipment (Chapter 3 of the guide)

Principle

Premises and equipment must be located, designed, constructed, adapted and maintained to suit the operations to be carried out. Their layout and design must aim to minimise the risk of errors and permit effective cleaning and maintenance in order to avoid cross-contamination, build up of dust or dirt and, in general, any adverse effect on the quality of products.

Premises

General

In summary:

- premises should be situated in an environment which, when considered together with measures to protect the manufacture, presents minimal risk of causing contamination of materials or products;

- premises should be carefully maintained, ensuring that repair and maintenance operations do not present any hazard to the quality of products. They should be cleaned and, where applicable, disinfected according to detailed written procedures;

- lighting, temperature, humidity and ventilation should be appropriate and such that they do not adversely affect, directly or indirectly, either the medicinal products during their manufacture and storage, or the accurate functioning of equipment;

- premises should be designed and equipped so as to afford maximum protection against the entry of insects or other animals;

- steps should be taken in order to prevent unauthorised people from coming in. Production, storage and quality control areas should not be used as a right of way by personnel who do not work in them.

Production Area

These should be designed to:

- minimise the risk of a serious medical hazard due to cross-contamination. Dedicated and self contained facilities must be available for the production of particular medicinal products, such as highly sensitising materials (eg penicillin) or biological preparations (eg from live micro-organisms);

- to allow the production to take place in areas connected in a logical order corresponding to the sequence of the operations and to the requisite cleanliness levels;

- provide adequate working and in-process storage space and should permit the orderly and logical positioning of equipment and materials so as to minimise the risk of confusion between different medicinal products or their components;

- be easily cleaned.

Storage Areas

These should be:

- of sufficient capacity to allow orderly storage of the various categories of materials and products and be designed or adapted to ensure good storage conditions;

- designed so that quarantine areas can be clearly marked and their access restricted to authorised personnel;

- designed to allow for segregation of rejected, recalled or returned materials or products;

- safe and secure areas.

Quality control areas

Normally:

- quality control laboratories should be separated from production areas. This is particularly important for laboratories for the control of biological, microbiological and radioisotopes, which should also be separated from each other;

- control laboratories should be designed to suit the operations to be carried out in them. Sufficient space should be given to avoid mid-ups and cross-contamination. There should be adequate suitable storage space for samples and records;

- separate rooms should be provided to protect sensitive instruments from vibration, electrical interference, humidity, etc;

- special requirements may be needed in laboratories handling particular substances, such as biological or radioactive samples.

Ancillary areas

These include:

- rest and refreshment rooms which should be separate from other areas;

- facilities for changing clothes, and for washing and toilet purposes and these should be easily accessible and appropriate for the number of users;

- maintenance workshops which should, as far as possible, be separated from production areas;

- animal houses which should be well isolated from other areas.

Equipment

The Guide specifies that:

- manufacturing equipment should be designed, located and maintained to suit its intended purpose;

- repair and maintenance operations should not present any hazard to the quality of the products;

- manufacturing equipment should be designed so that it can be easily and thoroughly cleaned;

- washing and cleaning equipment should be chosen and used in order not to be a source of contamination;

- equipment should be installed in such a way as to prevent any risk of error or of contamination. Production equipment should not present any hazard to the products;

- measuring, weighing, recording and control equipment should be calibrated and checked at defined intervals by appropriate methods. (Adequate records of such tests should be maintained);

- fixed pipework should be clearly labelled to indicate the contents and, where applicable, the direction of flow.

A11.7 Documentation (Chapter 4 of the Guide)

Principle

Good documentation constitutes an essential part of the quality assurance system. Clearly written documentation prevents errors from spoken communication and permits tracing of batch history. Specifications, Manufacturing Formulae and Instructions, Procedures, and Records must be free from errors and available in writing. The legibility of documents is of primordial importance.

Specifications describe, in detail, the requirements with which the products or materials used or obtained during manufacture have to conform. They serve as a basis for quality evaluation.

Manufacturing Formulae, Processing and Packaging Instructions state all the starting materials used and lay down all processing and packaging operations.

Procedures give directions for performing certain operations like cleaning, clothing, environmental control, sampling, testing, equipment operations.

Records provide a history of each batch of product, including its distribution, and also of all other relevant circumstances pertinent for the quality of the final product.

Documents should be:

- designed, prepared, reviewed and distributed with care;

- approved, signed and dated by appropriate and authorised persons;

- unambiguous;

- regularly reviewed and kept up-to-date;

- typed (or wordprocessed);

- amended and completed at the time each action is taken. They should be retained for at least one year after the expiry date of the finished product.

Note that data may be recorded by electronic data processing systems, photographic or other reliable means, but detailed procedures relating to the system in use should be available and the accuracy of the records should be checked. If documentation is handled by electronic data processing methods, only authorised persons should be able to enter or modify data in the computer and there should be a record of changes and deletions. Batch records electronically stored should be protected by back-up transfer on magnetic tape, microfilm, paper or other means. It is particularly important that, during the period of retention, the data are readily available.

Documents required

These include:

- specifications for starting and packaging materials;

- specifications for intermediate and bulk products;

- specifications for finished products;

- manufacturing formula and processing instructions;

- packaging instructions;

- batch processing and packaging records;

- procedures and records (receipt of materials, sampling, testing).

Here we will give just some of the details of this type of documentation.

Specifications for starting and packaging materials.

These should include:

- the designated name and the internal code reference;
- the reference, if any, to a pharmacopoeial monograph;
- the approved suppliers and, if possible, the original producer of the products;
- a specimen of printed materials;
    b) directions for sampling and testing or reference to procedures;

    c) qualitative and quantitative requirements with acceptance limits;

    d) storage conditions and precautions;

    e) the maximum period of storage before re-examination.

Specifications for finished products

These should include:

- the designated name of the product and the code reference where applicable;

- the formula or a reference to;

- a description of the pharmaceutical form and package details;

- directions for sampling and testing or a reference to procedures;

- the qualitative and quantitative requirements, with the acceptance limits;

- the storage conditions and precautions, where applicable;

- the shelf-life.

Manufacturing Formula and Processing Instructions

The Manufacturing Formula should include:

- the name of the product, with a product reference code relating to its specification;

- a description of the pharmaceutical form, strength of the product and batch size;

- a list of all starting materials to be used, with the amount of each, described using the designated name and a reference which is unique to that material; mention should be made of any substance that may disappear in the course of processing;

- a statement of the expected final yield with the acceptable limits, and of relevant intermediate yields, where applicable.

The processing Instructions should include:

- a statement of the processing location and the principal equipment to be used;

- the methods or reference to the methods, to be used for preparing the critical equipment (eg cleaning, assembling, calibrating, sterilising);

- detailed stepwise processing instructions (eg checks on materials, pretreatment, sequence for adding materials, mixing times, temperatures);

- the instructions for any in-process controls with their limits;

- where necessary, the requirements for bulk storage of the products;

- any special precautions to be observed.

Packaging instructions

These should normally include:

- name of the product;

- description of its pharmaceutical form, and strength where applicable;

- the pack size expressed in terms of the number, weight or volume of the product in the final container;

- a complete list of all the packaging materials required for a standard batch size, including quantities, sizes and types, with the code or reference number relating to the specifications of each packaging material;

- an example or reproduction of the relevant printed packaging materials and specimens indicating where to apply batch number references, and shelf life of the product;

- a description of the packaging operation, including any significant subsidiary operations, and equipment to be used;

- details of in-process controls and special precautions to be observed and acceptance limits.

Batch Processing Records

A Batch Processing Record should be kept for each batch processed. It should be based on the relevant parts of the currently approved Manufacturing Formula and Processing instructions.

During processing, the following information should be recorded at the time each action is taken and, after completion, the record should be dated and signed in agreement by the person responsible for the processing operations:

- the name of the product;

- dates and times of commencement, of significant intermediate stages and of completion of production;

- name of the person responsible for each stage of production;

- initials of the operator of different significant steps of production and, where appropriate, of the person who checked each of these operations (eg weighing);

- the batch number and/or analytical control number as well as the quantities of each starting material actually weighed (including the batch number and amount of any recovered or reprocessed material added);

- any relevant processing operation or event and major equipment used;

- a record of the in-process controls and the initials of the person(s) carrying them out, and the results obtained;

- the amount of product obtained at different and pertinent stages of manufacture (yield);

- notes on special problems including details, with signed authorization, for any deviation from the Manufacturing Formula and Processing Instructions.

Batch Packaging Records

These are analogous to the Batch Processing Records except they relate specifically to packaging.

Procedures and records

Receipt

There should be written procedures and records for the receipt of each delivery of each starting and primary and printed packaging material. The records of the receipts should include:

- the name of the material on the delivery note and the containers;

- the "in-house" name and/or code of material (if different from a);

- date of receipt;

- supplier's name and, if possible, manufacturer's name;

- manufacturer's batch or reference number;

- total quantity, and number of containers received;

- the batch number assigned after receipt;

- any relevant comment (eg state of the containers).

Sampling

- there should be written procedures for sampling.

Testing

- there should be written procedures for testing materials and products at different stages of manufacture.

Other

Written release and rejection procedures should be available for materials and products.

Records should be maintained of the distribution of each batch of a product in order to facilitate the recall of the batch if necessary.

A11.8 Production (Chapter 5 of the Guide)

Principle

Production operations must follow clearly defined procedures; they must comply with the principles of Good Manufacturing Practice in order to obtain products of the requisite quality and be in accordance with their manufacturing and marketing authorizations.

The overriding considerations are that;

- production should be performed and supervised by competent people;

- the handling of materials and products, such as receipt and quarantine, sampling, storage, labelling, dispensing, processing, packaging and distribution should be done in accordance with written procedures or instructions and, where necessary, recorded.

This latter consideration covers a wide range of issues. the Guide specifies many specific issues all aimed at ensuring quality. These include the quality of starting material the prevention of contamination and the validation of the product.

The Guide examines these under the heading of:

- general;

- prevention of cross-contamination in production;

- validation;

- starting materials;

- processing operations: intermediate and bulk products;

- packaging materials;

- packaging operations;

- finished products;

- rejected, recovered and returned materials.

A11.9 Quality Control (Chapter 6 of the Guide)

Principle

Quality Control is concerned with sampling, specifications and testing as well as the organisation, documentation and release procedures which ensure that the necessary and relevant tests are carried out, and that materials are not released for use, nor products released for sale or supply, until their quality has been judged satisfactory. Quality Control is not confined to laboratory operations, but must be involved in all decisions which may concern the quality of the product. The independence of Quality Control from Production is considered fundamental to the satisfactory operation of Quality Control.

The general conditions are that;

- each holder of a manufacturing authorization should have a Quality Control Department. This department should be independent from other departments, and under the authority of a person with appropriate qualifications and experience;

- finished product assessment should embrace all relevant factors, including production conditions, results of in-process testing, a review of manufacturing (including packaging) documentation, compliance with Finished Product Specification and examination of the final finished pack;

- Quality Control personnel should have access to production areas for sampling and investigation as appropriate;

- personnel, premises, and equipment in the laboratories should be appropriate to the tasks imposed by the nature and the scale of the manufacturing operations. The use of outside laboratories, can be accepted for particular reasons, but this should be stated in the Quality Control records.

Documentation should follow the principles given in Section A11.7. An important part of this documentation deals with Quality Control and the following details should be readily available to the Quality Control Department:

- specifications;
- sampling procedures;
- testing procedures and records (including analytical worksheets and/or laboratory notebooks);
- analytical reports and/or certificates;
- data from environmental monitoring, where required;
- validation records of test methods, where applicable;
- procedures for the records of the calibration of instruments and maintenance of equipment.

Any Quality Control documentation relating to a batch record should be retained for one year after the expiry date of the batch.

Other original data such as laboratory notebooks and/or records should be retained and readily available.

Sampling

Sample taking should be done in accordance with approved written procedures that describe:

- the method of sampling;
- the equipment to be used;
- the amount of the sample to be taken;
- instructions for any required sub-division f the sample;
- the type and condition of the sample container to be used;
- any special precautions to be observed, especially with regard to the sampling of sterile or noxious materials;
- the storage conditions;
- instructions for the cleaning and storage of sampling equipment.

Reference samples should be representative of the batch of materials from which they are taken. Samples taken from each batch of finished products should be retained until one year after the expiry date.

Testing

The general conditions of testing are that:

- analytical methods should be validated;

- the results obtained should be recorded and checked to make sure that they are consistent with each another;

- the tests performed should be recorded and the records should include at least the following data:

    a) name of the material or product and, where applicable, dosage form;

    b) batch number and, where appropriate, the manufacturer and/or supplier;

    c) references to the relevant specifications and testing procedures;

    d) test results, including observations and calculations, and reference to any certificates of analysis;

    e) dates of testing;

    f) initial of the persons who performed the testing;

    g) initials of the persons who verified the testing and the calculations, where appropriate;

    h) a clear statement of release or rejection (or other status decision) and the dated signature of the designated responsible person.

(*) In Germany, France and Greece, samples of starting materials should be retained for as long as the corresponding finished product.

- all the in-process controls, should be done according to methods approved by Quality Control and the results recorded;

- special attention should be given to the quality of laboratory reagents, volumetric glassware and solutions, reference standards and culture media. They should be prepared in accordance with written procedures;

- laboratory reagents intended for prolonged use should be marked with the preparation and expiry date and the signature of the person who prepared them;

- animals used for testing components, materials or products, should, where appropriate, be quarantined before use.

A11.10 Contract Manufacture and Analysis (Chapter 7 of the Guide)

Principle

Contract manufacture and analysis must be correctly defined, agreed and controlled in order to avoid misunderstandings which could result in a product or work of unsatisfactory quality. There must be a written contract between the Contract Giver and the Contract Acceptor which clearly establishes the duties of each part. The contract must clearly state the way in which the Qualified Person releasing each batch of product for sale exercises his full responsibility.

Note: This Chapter deals with the responsibilities of manufacturers towards the Competent Authorities of the Member States with respect to the granting of marketing and manufacturing authorisations It is not intended in any way to affect the respective liability of contract acceptors and contract givers to consumers; this is governed by other provisions of community and national law.

The general conditions which are applied are that;

- there should be a written contract covering the manufacture and/or analysis arranged under contract and any technical arrangements made in connection with it;

- all arrangements for contract manufacture and analysis including any proposed changes in technical or other arrangements should be in accordance with the marketing authorization for the product concerned.

The Contract Giver and Acceptor

The Guide provides detailed descriptions of responsibilities of the contract giver and the contract acceptor and outlines the types of specifications that should be included in the contract.

A11.11 Complaints and product recall (Chapter 8 of the Guide)

Principle

All complaints and other information concerning potentially defective products must be carefully reviewed according to written procedures. In order to provide for all contingencies, a system should be designed to recall, if necessary, promptly and effectively products known or suspected to be defective from the market.

The Guide specifies that an individual must hold the responsibility for handling complaints and for deciding the measures that need to be taken. Such measures should follow written procedures describing the action(s) to be taken. The Guide also provides the framework for conducting the recall of suspect materials and products.

A11.12 Self Inspection (Chapter 9 of the Guide)

Principle

Self inspections should be conducted in order to monitor the implementation and the respect of Good Manufacturing Practice principles and to propose necessary corrective measures.

The general conditions are:

- personnel matters, premises, equipment, documentation, production, quality control, distribution of the medicinal products, arrangements for dealing with complaints and recalls, and self inspection, should be examined at intervals following a pre-arranged programme in order to verify their conformity with the principles of Quality Assurance;

- self inspections should be conducted in an independent and detailed way by designated competent person(s) from the company. Independent audits by external experts may also be useful;

- all self inspections should be recorded. Reports should contain all the observations made during the inspections and, where applicable, proposals for corrective measures. Statements on the actions subsequently taken should also be recorded.

A11.13 Supplementary Guidelines. Manufacture of Sterile Medicinal Products

Principle

Manufacture of sterile preparations needs special requirements to minimize risks of microbiological contamination, and of particulate and pyrogen contamination. Much depends on the skill, training and attitudes of the personnel involved. Quality Assurance bears a particularly great importance, and this manufacture must strictly follow carefully established and validated methods of preparation.

The Guide includes 85 conditions that should be applied to the manufacture of sterile medicinal products. These cover general requirements concerning the premises in which such manufacture takes place, conditions which apply to personnel engaged in the manufacture, the quality and maintenance of equipment and the activities undertaken during the manufacturing process. Particular attention is directed towards sterilisation processes including sterilisation by heat radiation, ethylene oxide and filtration.

The Guide divides manufacturing operations into two categories;

- those in which the preparation is terminally sterilised, sealed in its final container;

- those in which aseptic conditions are necessary at some (or all) stages of the manufacturing processes.

The general conditions which apply are that:

- production of sterile preparations should be carried out in clean areas whose entry should be through airlocks for personnel or for goods. Clean areas should be maintained to an appropriate cleanliness standard and supplied with air which has all passed through filters of an appropriate efficiency;

- the various operations of component preparation, product preparation, filling and sterilisation should be carried out in separate areas within the clean area;

- each manufacturing operation requires an appropriate air cleanliness level in order to minimise the risks of particulate or microbial contamination of the product or materials being handled. The Guides give detailed minimum air-grades required for different manufacturing operations. It also gives guidance as to how these air standards may be achieved. It also points out the utilisation of absolute barrier technology and automated systems to minimise human interventions in processing areas can produce significant advantages in assurance of sterility of manufactured products.

Regulation of biotechnology in the food industry

Regulation of biotechnology in the food industry

12.1 Introduction

Regulations governing the production and marketing of foods have a long and complex history. In former times, these regulations were generated and applied nationally. A great diversity of measures were implemented and a variable range of standards used. Latterly the increasing internationisation of trade in food stuffs has driven the development of more commonly accepted standards and the production of mutually acceptable practices. This is especially reflected by developments within the EC where a long series of Directives and Communications provide the basis for the harmonisation of standards and procedures. The effect of the single European Act (1986) has been to speed up this process of harmonisation. This is, in principle, beneficial and will do much to remove barriers to trade between Member States. Furthermore, the adoption of similar standards and procedures by the Food and Drug Administration (FDA) in the USA will also facilitate transatlantic trade.

Food law is extensive and covers many issues ranging from the production and manufacture of food to its marketing. A list of aspects of food law is provided in Table 12.1.

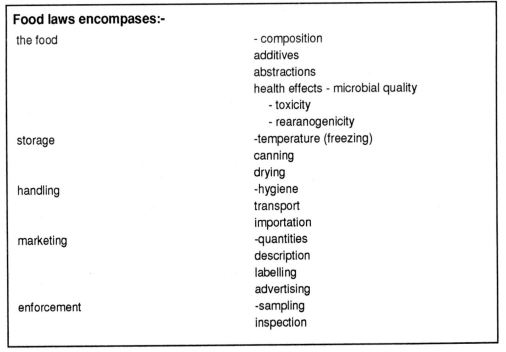

Food laws encompases:-	
the food	- composition
	additives
	abstractions
	health effects - microbial quality
	- toxicity
	- rearanogenicity
storage	-temperature (freezing)
	canning
	drying
handling	-hygiene
	transport
	importation
marketing	-quantities
	description
	labelling
	advertising
enforcement	-sampling
	inspection

Table 12.1 Examples of subjects encompassed within Food Regulations.

The information included in this table illustrates how extensive Regulations governing the production, distribution and sale of food are.

The regulations encompass three main principles:

- the food may not be dangerous to public health;

- the food may not impede fair trade;

- the food (or its marketing) may not mislead the consumer.

The Regulations are therefore, concerned with protecting the consumer as well as the producer/trader.

It is not our intention here to review the whole range of these Regulations. We will predominantly focus onto those Regulations which impinge on the application of contemporary biotechnology in the food industry. We will begin by briefly describing the role of contemporary biotechnology in food production and manufacture and then we will consider the main EC-Regulations that impinge upon these applications. It should be realised, however, that there are some similarities between the regulation of food production and marketing and the manufacture and use of medicines: both food and medicines are, after all, destined for similar 'internal' use.

If you require further details on Food Law, we recommend A A Painter's text 'Butterworths Food Law, 1992' Butterworths, London ISBN 0406 00642 3. This text, written mainly from a UK perspective, provides an extensive review of EC legistration. Of particular value is the Digest of the Principal Instruments of European Community Food Law presented as an appendix. This appendix covers the following range of products and issues.

additives; alcoholic drinks;caseins and
casinates;   chestnut puree;
articles appearing in food; Cocoa   labelling
and chocolate products   materials and articles in contact with food;
coffee extracts and chicory extracts;   meat
eggs;   meat products
enforcement and administration   poultry and poultry meat
foodstuffs for particular   product liability
fresh fruit and vegetables;   radioactive contamination
fruit juices and related products;   sugar
honey;   quick frozen foodstuffs
jams, jellies, and marmalades   wine.

It also provides lists of permitted additives (antioxidants, colours emulsifiers and stabilisers, mineral hydrocarbons, preservative and other additives). Also important are some useful addresses.

12.2 Contemporary biotechnology and food production

The contribution of contemporary biotechnology to food production and manufacture can be divided into a number of different groups of activities. These are:

- measures which lead to increased crop production (for example genetically modifying plants to improve yield by increasing disease resistance, treatment with biocides, etc);

- measures which lead to increased animal yields (for example vaccination, growth hormone treatment);

- new routes for the manufacture of 'traditional' food additives (for example amino acids, vitamins, preservatives, colourants, enzymes);

- new (modified) organisms to bring about food transformations (for example in the manufacture of cheese and yoghurt);

- production of completely novel foods and food ingredients.

All of these are of course subject to many of the same regulations which control the manufacture of conventional foods but there are some important special considerations.

12.3 Types of regulations and guidelines

At both national and supranational levels, we can distinguish various types of regulations. Basically we can differentiate between legislations (for example, EC-Directives, Member State laws) issued by a body with legislative power for example the (EC) and guidelines issued by international organisation (for example FAO/WHO-JECFA, AMFEP, OECD- see below) which do not have legal authority.

(FAO = Food and Agriculture Organisation; WHO = World Health Organisation, JECFA = Joint Expert Committee on Food Additives; AMFEP = Association of Microbial Food Enzyme Producers, OECD = Organisation for Economic Co-operation and Development).

Within the EC, a large number of Directives exist that controls, for example, additives to food or the amount of residues (for example, pesticides used in crop protection, hormones used in enhancing growth). Such Directives apply to additives and/or residues made by whatever route. None specifically refer to biotechnological products.

In Table 12.2 we have provided examples of Directives relating to residues remaining in food whilst Table 12.3, lists EC Directives which deal with food additives.

> **Pesticide Residues**
>
> Directive 86/362 EEC (OJ L221; 07.08.86)
> Directive 86/362 EEC OJ L221; 07.08.86)
> dealing with the maximum levels of pesticides residues allowable for cereals and for foods derived from animals - These Directives have not been fully implemented. Instead, the residue limits recommended by Codex (Codex Alimentainus Commission) have been used. Directive 90/642/EEC (OJ L350: 14:12:90) A framework Directive for residue limits for fruit and vegetables. This will provide the framework for the new residue limits for there production.
>
> **Veterinary Residues**
>
> Directives 81/602/EEC (OJ L 222;07.08.81)
> 86/469/EEC (OJ L 275;26.09.86)
> 88/146/EEC (OJ L070; 16.03.88)
> 85/358/EEC (OJ L191; 23.07.85)
>
> - These prohibit the use of certain substance (growth promoters) and restrict the use of others. Do not include farmed fish and poultry.

Table 12.2 EC-Directives dealing with residues in food.

guidelines of international organisations

In contrast to EC-Directives, guidelines issued by international organisations concerning food additives and enzymes are usually formulated as specifications. Their particular aim is to guarantee the safety of the preparation. For example, the Joint Expert Committee on Food Additives (JECFA) of FAO and WHO has issued general microbiological and chemical specifications for the identity and purity' of enzyme preparations (JECFA WHO/FAO "Specification for identity and purity 1981 Vol 19 p214-216). These specifications are widely accepted by regulatory authorities all over the world and are considered to be of fundamental importance in the safety evaluation process of enzyme preparations. An analogous set of guidelines has been issued by the Association of Microbial Food Enzyme Producers (AMFEP) (see for example AMFEP 'Regulatory aspects of microbial food enzymes' 1988).

horizontal and vertical regulations

Before we move on to discuss the actual processes involved in getting a new product, or an 'old' product made by a new route, to the market place, we would also like to make the point that we can also identify two sub-groups of legislation and guidelines. Some may be regarded as 'horizontal' that is they refer to many different food types. For example, regulations referring to preservatives and emulsifiers may involve many different foodstuffs. Other regulations and guidelines may, however, refer to the production of a particular food type. Such regulations are often referred to as 'vertical' regulations. Thus, in placing a food or food ingredient on the market we may have to satisfy both 'vertical' and 'horizontal' regulations.

Let us assume that you have devised a new food or food ingredient. How do you gain approval (authorisation) to market your product? We will deal with this in the following section.

12.4 The registration (authorisation) procedure

For new products or for products made by new routes, a registration process analogous to that imposed on the marketing of medicines is followed.

General

Directive 89/107/EEC - A framework Directive on the control of additives (OJL40: 11-02-89)
Directive 88/788/EEC A framework Directive on the control of flavourings and completed by
Directive 91/71/EEC (OJ. L42: 15.02.91)

Antioxidants

Directive 70/357/EEC - Control and positive list of approved antioxidants (OJ L157:
18.07.70)

(Amended by	73/101/EEC	OJ L022-01.0.1.73
	74/412/EEC	OJ L221 12.08.74
	78/143/EEC	OJ L044 15.02.78
	81/962/EEC	OJ L354 09.12.81
	85/007/EEC	OJ L002 03.01.85
	87/055/EEC	OJ L0234 27.01.87)

Directive 78/664/EEC - purity criteria (OJ L186: 30.06.78)

Colours

Directive 2545/62/EEC Control and positi ve list of food colours (OJ L115:11.11.62)
(Amended and supplemented by Directives 65/469;70/358; 73/101; 76/399; 78/144;
81/712; 85/007; 67/653; 68/419;81/020)

Emulsifiers, Stabilisers, Thickners and Gelling Agents

Directive 74/329/EEC - Control and positive lists of emulsifiers,, stabilisers, thickeners and
gelling agents (OJ L 189, 12. 07.74
(Amended by Directive 78/612; 80/597; 85/006; 85/007; 86/102; 898/393) 898/393)
Decision 78/663/EEC - purity criteria for emulsifiers, stabilisers, thickness and gelling
agents. (OJ 2230; 05.08.82)
Directive 90/612) - purity criteria for emulsifiers etc. (OJ L326. 24.11.90)

Preservatives

Directive 64/54/EEC - control and psitive list of preservatives (OJ L012; 27.01.64)
(Ammended by Directives 65/569; 66/722; 67/427; 68/420; 70/359; 71/160; 72/002; 72/002;
72/444; 73/101; 74/062; 74/394; 76/462; 74/394; 76/462, 78/145; 81/214; 85/007; 85/172/
85/585.)
Directive 65/66/EEC - purity criteria for preservatives (OJ L 022; 09.02.65)
(Amended by Directive 67/428; 76/428; 76/463; 86/604.)

Others

Extraction solvents - Directive 88/344/EEC - list of extraction solvents which may be used
in food and restrictions on use (OJ L157; 24.06.88)
Erucic acid- Directive 76/621/EEC - restriction of erucic and content of oil or fat or food
supplemented with oil or fat (OJ L157; 28.07.76)

Table 12.3 EC- Directives dealing with food additives (data from Painter AA. 1992 Butterworths Food Law,
Butterworths London.

In applying for registration to market a product, the applicant first needs to establish
the regulatory requirements for the authorisation to market a product in target
countries. It is best to make such inquiries directly to the relevant national authorities
although industrial organisations, universities and independent institutes may also be
used. It is usual to conduct a dual approach to establishing the safety of the product.
These are:

- through extensive toxicological testing of the product and a detailed
  characterisation of the production process;

- through establishing the similarity of the product with natural occurring product(s) or products made by alternative routes.

It is also important to establish the efficacy of the product. Although not always defined as a requirement for obtaining authorisation, it is vital in order to get the industrial support to prove to the authorities that the product is needed.

Trial programmes are usually carried out in target countries using independent organisations (for example Universities).

The nature of the safety evaluation programme is, of course dependent upon the product and the processes involved in its production. However, the toxicological tests are similar to those carried out with products made by conventional methods.

Often the products made by contemporary biotechnology depend upon the use of genetically modified organisms. Thus, inevitably within the EC, the EC-Directives on genetically modified organisms (see Chapter 6) apply. These Directives do not specify the nature of the use of the product. In the manufacture of food additives and enzymes using such organisms, if the production and recovery processes are such that no living recombinant organisms are present in the final product no further measures are necessary to reduce or destroy the environmental fitness of the organism. It is however highly recommended to avoid the use of any markers that code for resistance to antibiotics that are in current use in human or vetinary medicine. This is to avoid further spreading of resistance to these antibiotics. If however such markers are used precautions need to be taken to prevent DNA carrying such genes from being present in the product.

In Table 12.4 we have summarised a typical example of the contents of a safety dossier which would be submitted to the appropriate authorities to gain the registration of a process to make a food additive/enzyme using a genetically modified micro-organism.

In some cases however, whole organisms remain (or may remain) in the product. For example the manufacture of yoghurt and cheese involve the cultivation of selected micro-organisms in the food and the organisms themselves become an integral part of the food. The organisms traditionally used in these processes are only those which have a long history of safe use. However the advent of genetic engineering has enabled food processors to generate organisms of greater industrial value. Here again, the safety of the final product is of paramount importance and the onus is on the manufacturer of the product to prove that this genetic insert modification is not likely to lead to any new problems of product safety.

Part I	Production organism	
		Pathogenicity
		Stability of DNA
		Characterisation of DNA
Part II	Product Specifications	
		Physicochemical Characterisation
		Microbiological Specifications
		Chemical Specifications
Part III	Product Toxicology	
		Short term studies
		Subchronic study
		Mutagenicity studies
		Allergenicity study
Part IV	Product Efficacy	

Table 12.4 The contents of a typical safety dossier for a food product.

12.5 Novel foods and novel food ingredients

The EC has recently issued (July 1992) a proposal for a Council Regulation on novel foods and novel food ingredients (92/C 190/04/EEC - OJ C193/3 .29.7.92) This proposal is quite brief and is presented in its entirety below. It came into force on 1st January, 1993.

As with most EC-regulations, a number of exemptions are specified. The proposal does, however, indicate the relationship between novel foods and the EC-Directive (90/220/EEC) which controls the release of genetically modified organisms (see Chapter 6).

The proposal indicates that Member States will establish a list of independent experts qualified to carry out specified examinations. The proposer must provide the competent authority with evidence of the safety and efficacy of the product. The registration of the product may specify the condition of use of the food or food ingredient and may also establish a name for the product and indicate the form of labelling used.

The proposals also makes provision for the removal of the product from the market (Article 9) if it is thought that the product may endanger human health.

This Directive is of fundamental importance to contemporary biotechnology so it is worthwhile examining these regulations in detail.

COMMISSION

Proposal for a Council Regulation (EEC) on novel foods and novel food ingredients

(92/C 190/04)

COM(92) 295 final - SYN 426

(Presented by the Commission on 7 July 1992)

THE COUNCIL OF THE EUROPEAN COMMUNITIES,

Having regard to the Treaty establishing the European Economic Community and in particular Article 100a thereof,

Having regard to the proposal from the Commission,

In co-operation with the European Parliament,

Having regard to the opinion of the Economic and Social Committee,

Whereas differences between national laws relating to novel foods or food ingredients could hinder the free movement of foodstuffs; whereas they may create conditions of unfair competition, thereby directly affecting the establishment or functioning of the internal market

Whereas the measures aimed at the gradual establishment of the internal market must be adopted by 31 December 1992 whereas the internal market consists of an area without internal frontiers within which the free movement of goods, persons, services and capital is guaranteed;

Whereas, also, the smooth running of the internal market requires that provisions for notification and authorisation of foods or food ingredients which have not hitherto been used for human consumption to a significant degree in the Community and/or when have been produced by food production processes that result in a significant change in their composition and/or nutritional value and/or intended use should be determined at Community level; whereas such authorisation shall be of general application;

Whereas this Regulation does not affect food additives, flavourings for use in foodstuffs and extraction solvents falling within the scope of other Community provisions;

Whereas risks to the environment may be associated with food or food ingredients which contain or consist of genetically modified organisms; whereas Directive 90/220/EEC has specified that, for such products, an environmental risk assessment must always be undertaken to ensure safety for the environment; whereas, in order to provide for a unified Community system for assessment of a product, provisions must be made under this regulation for a specific environmental risk assessment, which in accordance with the procedures of Article 10 of Directive 90/220/EEC must be similar to that laid down in that Directive, together with the assessment of the suitability of the product to be used as a food or food ingredient;

Whereas the Scientific Committee for Food has to be consulted on any decision on foods or food ingredients which have not hitherto been used for human consumption to as significant degree in the Community and/or which have been produced by food production processes that result in a significant change in their composition and/or nutritional value and/or intended use likely to have an effect on public health;

Whereas, in respect of this Regulation, provision should be made for a procedure instituting close cooperation between Member States and the Commission within the Standing Committee on Foodstuffs set up by Council Decision 69/414/EEC(1),

HAS ADOPTED THIS REGULATION:

Article 1

This Regulation lays down provisions for the placing on the market of foods or food ingredients which have not hitherto been/used for human consumption to a significant degree/and/or which have been produced by processes that result in a significant change in their composition and co nutritional value and/or intended use. The categories of products falling within the scope of this Regulation are listed in Annex I.

Article 2

This Regulation shall not apply to:

a) food additives falling within the scope of council Directive 89/107/EEC(1);

b) flavourings for use in foodstuffs, falling within the scope of Council Directive 88/388/EEC (2);

c) extraction solvents used in the production of foodstuffs, falling within the scope of Council Directive 88/344/EEC(3);

d) foods and food ingredients treated with ionizing radiation, falling within the scope of Council Directive../.../EEC.

Article 3

1. Member states shall establish a list of independent experts with scientific experience qualified to carry out the examinations referred to in Article 5.

2. The list and subsequent modifications shall be notified to the Commission. The Commission shall compile the consolidated list of experts from the notifications and shall ensure its publication.

3. Criteria for the selection of the experts mentioned in paragraph 1 may be adopted in accordance with the procedure laid down in Article 10.

Article 4

1. A food or food ingredient falling within the scope of this Regulation shall be placed on the market for the first time in accordance with the procedure stipulated in Article 5; however, where the food is consumed as a viable organism or where generally accepted scientific data are not available to demonstrate its safety it is submitted to the procedure of Article 6.

Article 5

1. Where on the basis of generally accepted scientific data, in the opinion of one or more of the qualified experts from the list mentioned in Article 3, there is evidence that the product to be used as a food or food ingredient complies with the general criteria mentioned in Anex II, the person legally responsible shall notify the Commission with a summary of the evidence together with the opinion of the expert.

The Commission shall immediately send the notification to the Member States.

2. The food or food ingredient concerned may be placed on the market only three months after the notification received by the Commission and provided the Commission has not delivered a negative opinion within the period of three months at its own initiative or at the duly motivated request from a Member State.

In the event of a negative opinion, the procedure in Article 6 is to be followed.

3. For the purposes of control, where necessary, the competent authority shall be empowered to

require the persons legally responsible for placing the product on the market to produce the scientific work and the data establishing the product's compliance with the general criteria mentioned in Annex II and with the procedure laid down in Article 10.

Article 6

1. When the procedure laid down in this Article is to be followed, the person legally responsible for placing the product on the market in the Community shall submit a request for authorization to the Commission comprising the necessary information to assure the compliance with the criteria mentioned in Annex II. The Commission shall inform the Member States accordingly. The Member States may send to the Commission observations including pertinent scientific information.

2. A decision shall be taken on the authorization for the marketing of the food or food ingredient according to the procedure laid down in Article 10.

3. The decision mentioned in paragraph 2 may establish the conditions of use of the food or food ingredient when appropriate. It may also establish the name of the food or food ingredient and any indications concerning the labelling, as the case may be, as laid down in Article 5 of Council Directive 79112/EEC on the approximation of the laws of the Member States relating to the labelling, presentation and advertising of foodstuffs for sale to the ultimate consumer(1)

4. The Commission shall communicate to the applicant the decision taken with respect to his request.

5. The Commission shall inform Member States of any decision adopted pursuant to paragraph 2.

6. Detailed rules for implementing this article may be adopted in accordance with the procedure laid down in Article 10.

Article 7

1. Where the food or food ingredient falling within the scope of this Regulation contains or consists of a genetically modified organism within the meaning of Article 2 (1) and (2) of Council Directive 90/220/EEC (2) on the deliberate release of genetically modified organisms, the information required in the request for authorization mentioned in Article 6 shall be accompanied by:

a copy of the written consent, from the competent authority, to the deliberate release of the genetically modified organisms for research and development purposes provided for in Article 6 (4) of Directive 90/220/EEC, together with the results of the release(s) with respect to any risk to human health and the environment,

the complete technical dossier supplying the information requested in Annexes II and III of Directive 90/220/EEC and the environmental risk assessment resulting from this information.

Articles 11 to 18 of Directive 96/220/EEC shall not apply to food or food ingredients falling within the scope of Article 6 which contain or consist of a genetically modified organism.

2. In the case of food or food ingredients falling within the scope of this Regulation containing or consisting of a genetically modified organism, the decision mentioned in Article 6 (2) shall take account of the environmental safety requirements laid down by Directive 90/220/EEC

3. Detailed rules for implementing this Article may be adopted in accordance with the procedure laid down in Article 10.

Article 8

Any decision or provision regarding food or food ingredient falling within the scope of Article 1 likely to have an effect on public health shall be adopted by the Commission after consultation with the Scientific Committee for Food, either on its own initiative or at the request of a Member State.

Article 9

1. Where a Member State had detailed grounds for considering that the use of a food or a food ingredient falling within the scope of Article 1, although it complies with this Regulation, endangers human health, that Member State may temporarily suspend or restrict the trade and use of the food or food ingredient in question in its territory. It shall immediately inform the other Member States and the Commission thereof and give reasons for its decision.

2. The Commission shall examine the grounds given by the Member State referred to in paragraph 1 as soon as possible within the Standing Committee for Food stuffs, and shall then deliver its opinion forth with and take the appropriate measures following the procedure laid down in Article 10.

3. If the Commission considers that the national measure must be dispensed with or modified, it shall initiate the procedure laid down in Article 10 for the adoption of the appropriate measures.

Article 10

1. Where the procedure laid down in this Article is to be followed, the Commission shall be assisted by the Standing Committee on Foodstuffs, set up pursuant to Decision 69/414/EEC(3) acting in an advisory capacity, hereafter referred to as 'the Committee'.

2. The Chairman shall submit to the Committee a draft of measures to be taken. The Committee shall deliver its opinion on the draft within a time limit which the Chairman may lay down according to the urgency of the matter, if necessary by taking a vote.

3. The opinion shall be recorded in the minutes; in addition, each Member State shall have the right to ask to have its position recorded in the minutes.

4. The Commission shall take the utmost account of the opinion delivered by the Committee. It shall inform the Committee of the manner in which its opinion has been taken into account.

Article 11

This Regulation shall enter into force in 1. January 1993.

This Regulation shall be binding in its entirety and directly applicable in all Member States.

ANNEX I

Categories of products falling within the scope of this Regulation:

- a product *consisting* of or containing a modified food molecular entity, or a molecular entity with *no established history of food use*;

- a product produced from, or consisting of, or containing an organism, or part of an organism, *currently* used in food production which has been modified by gene technology;

- a product to which has been applied a process not currently used for food manufacture or which, although subjected to such a process, has not previously been placed on the market and where such a process gives rise to significant changes in the composition or structure of the end product which affect its nutritional value and/ or its digestibility and/or its metabolism and/or level of undesirable substances in the food.

ANNEX II

- General criteria for the placing on the market of foods and food ingredients falling within the scope of Article 1;

- Foods and food ingredients falling under the scope of Article 1 may be placed on the market provided that.

12.6 Additional comments

In the previous sections, we have indicated that a variety of regulations and guidelines must be adhered to in order to gain market authorization for new food materials or new processes to make such products. Obviously good laboratory/manufacturing practices, and described in earlier chapters must be adhered to. Similarly high hygiene standards must be maintained. It must however be made clear that Food regulations are not merely confined to the manufacturer of the food, its food value and safety. It also involves a wide variety of measures such as labelling and advertising. We will illustrate this using a 'typical' conventional biotechnology product, wine. The wine industry is a traditional industry where each wine has its own particular characteristic. Many legal measures are taken to protect the consumer. Many of these measures are concerned not directly with consumer safety but to ensure consumer is not cheated by incorrect labelling. They also protect producers from counterfeit versions of their products and guard against unspecified adulteration of products. Table 12.5 summarises the main instruments of EC Food Law specifically related to wines. Remember that, in addition to these, a wide number of more generalised Directives also apply.

It should be self-evident from this single example that the legal requirements that need to be met for marketing food products is a specialist area and requires specialist expertise. Of course, not all products are as complex as wine.

Analysis

Regulation 2676/90/EEC Community methods of analysis of wine products (OJ L272, 03.10.90)

Aromatized wines

Regulation 1601/91/EEC General rules for aromatized wines, aromatized wine based drinks and aromatized wine products cocktails (OJ L149; 14.06.91)

Regulation 366e/91/EEC Transitional measures relevant to 1601/91/EEC

Description

a) wine	2391/89/EEC;	OJ. L232 9.08.89
Regulation	3886/89/EEC	OJ .L378 27.12.89
	3827/90/EEC	OJ. L366 29.12.90
	2356/91/EEC	OJ. L216 03.08.91

b) wine/grape musks

Regulation 3201 90 Detailed rule for the description of wine and grape musks
(OJ L 309. 08.11.90)

c) sparkling wine

Regulation 3309/85/EEC Description of sparkling wines and aerated sparkling wines.
(OJ L 320 29.11.85)

(Amended by	3805/85/EEC	OJ L367, 31.12.85
	1626/86/EEC	OJ L144, 29.05.86
	0538/87/EEC	OJ L055, 25.02.87
	2045/89/EEC	OJ L202, 14.07.89
	2357/91/EEC	OJ L214, 03.08.91)

Regulation 2707/86/EEC Detailed rules for sparkling and aerated sparkling wines (OJ L 246, 30.08.86)

Amended by	3778/86/EEC	OJ L319 5.11.86
	2249/87/EEC	OJ L207 29.07.87
	0575/88/EEC	OJ L237 02.03.88
	2657/88/EEC	OJ L237 27.08.88
	0596/89/EEC	OJ L065 09.03.89
	2776/90/EEC	OJ L267 29.09.90
	3826/90/EEC	OJ L366 29.12.90

Enriching and acidification

Regulation 2094/86/EEC. Rules for the use of tartaric acid for deacidification of certain wines products by (OJ L180 04.07.86)

(Amended by	2736/86/EEC	OJ L252. 04.09.86)

Regulation 2240/89/EEC Monitoring the use of enriching and deacidification of wine (OJ L 215 260789)

Organisation of the Market Regulations

	822/87/EEC	OJ L084 27.03.87
	1390/87/EEC	OJ L133 22.05.87
	1972/87/EEC	OJ L184 03.07.87
	3146/87/EEC	OJ L300 23.10.87
	3992/87/EEC	OJ L377 31.12.87
	1441/88/EEC	OJ L132 28.05.88

continued

Table 12.5 The main EC-Regulations which specifically apply to wine (Data from Painter AA. Butterworth's Food Law 1992. Butterworth's London). (continued)

	2253/88/EEC	OJ L198 26.07.88
	2964/88/EEC	OJ L269 29.09.88
	4252/88/EEC	OJ L373 31.12.88
	1236/89/EEC	OJ L128 11.05.89
	0388/90/EEC	OJ L042 06.02.90
	1325/90/EEC	OJ L132 23.05.90
	3577/90/EEC	OJ L148 13.06.91
	1734/91/EEC	OJ L163 26.06.91

Quality Wines

Control of:

Regulation 2046/89/EEC General rules on control
(OJ L20214.07.89)

Regulation 1698/70/EEC - On certain derogations on
production (OJL190. 2608.70)

(Amended by	0101/73/EEC	OJ L 002 01.01.73
	0807/73/EEC	OJ L078 27.03.73
	0986/89/EEC	OJ L106 18.04.89)

Regulation 2247/73/EEC - on quality wines produced in specifiedregions (OJ L230 18.08.73)

(Amended by	0418/86/EEC	OJ L048 26.02.86
	0986/89/EEC	OJ L106 18.04.89)
	8231/87/EEC	OJ L084 27.03.87)

Regulation 3590/83 method of analysis for alcohol
(OJ L363: 24.12.83)

(Amended by	1614/86/EEC	OJ L142 28.05.86
	2039/88/EEC	OJ L179 09.07.88)

Regulation 4252/88/EEC Preparation of liqueur wines (OJ L373. 31.12.88)

(Amended	1328/90/EEC	OJ L132 23.05.90
	1735/91/EEC	OJ L163 26.06.91)

List of approved:

Regulation 2805/73.EEC List of quality white wines
(OJ L230 18.08.73)

(Amended by	3548/73/EEC	OJ L361 29.12.73
	2160/75/EEC	OJ L220 20.08.75
	0966/77/EEC	)OJ L115 06.06.77

Regulation 2082/74/EEC List of quality liqueur wine produced in specified regions. (OJ L217 08.08.74)

Coupage:

Regulation 479/86/EEC Coupage of red Spanish wines with red wines from other Member States

Regulation 1781/86/EEC Rules of coupage with Spanish red wines

Regulation 643/77/EEC Wines from other countries
(Amended by 3203/80/EEC and 0418/86/EEC

Prescribed forms.

Regulation 1153/75/EEC Wine producer and traders
(Amended by 2617/77/EEC; 3203/80/EEC; 0418/86/EEC; 0986/89/EEC.)   Continued

Table 12.5 The main EC-Regulation which specifically apply to wine (Data from Painter AA. Butterworth's Food Law 1992. Butterworth's London).

Regulation 986/89/EEC Documents to accompany carriage of wine and record to be kept (Amended by 2245/90/EEC)

Maximum alcoholic strength

Regulations 319/74/EEC Wine growing areas which may produce table wines having a prescribed maximum alcoholic strength (OJ L248 11.09.74)

Regulation 351/79/EEC on the addition of alcohol to wines (OJ L054 05.03.79)

(Amended by		
	3658/81/EEC	OJ L 366 22.12.81
	0255/87/EEC	OJ L 026 29.01.87
	4090/87/EEC	OJ L 382 31.12.87
	3904/88/EEC	OJ L 342 16.12.88
	1372/90/EEC	OJ L 133 24.05.90

Measurement indication of alcoholic strength

Regulation 1627/86/EEC - General rules (OJ L144 29.05.86)

Regulation 1069/87/EEC Indication on special wine (OJ L 104 16.04.87)

Regulation 3677/87/89 Total alcohol/audity of imported wine (OJ L360 09.12.89)

Sparkling wines

Regulation 2152/75/EEC Detailed rules (OJ L219 19.08.75)
                    (Amended 0986/89/EEC OJ L 106 18.04.89)

Regulation 358/79/EEC Community produced wines (OJ L054, 05.03.79)

(subsequent amended by   2383/79/EEC, 3456/80/EEC;3686/84/EEC, 3310/85/EEC, 3805/86/EEC 2044/89/EEC)

Imports

Regulation 1393/76/EEC - Detailed rules on importation from third countries. OJ L157 18.06.86

(Amended by   0688/78/EEC; 1666/78/EEC; 1243/79/EEC;3104/80/EEC, 3671/89/EEC.

Regulation 2390/89. General rules for importation of wines, grape juice and grape must. (03. L232. 09.12.89)

Amended by   3887/87/EEC, 2179/90/EEC, 331/91/EEC

Table 12.5 The main EC-Regulations which specifically apply to wine (Data from Painter AA. Butterworth's Food Law 1992. Butterworth's London). (continued)

A general comment on the biotechnological production of chemicals other than medicines and food ingredients

A general comment on the biotechnological production of chemicals other than medicines and food ingredients

In Chapters 10-12, we have commented upon the production of materials (chemicals) for use in the medical and food industries. These are, of course, the main product areas in which biotechnology finds application. We have also indicated that the release of genetically modified organisms for, agricultural use is also controlled by EC-Directives. However, the prospects are increasing for using biotechnology to produce specific chemicals which have applications other than as medicines and foods. Here we point out that it is the responsibility of the manufacturer (or importer) of a new substance to notify competent authorities (in UK, Health and Safety Executive and the Department of the Environment) before marketing it. This responsibility is specified by the EC 'Dangerous Substances Directive'). Data concerning the product must be provided to enable the authority to assess its potential hazard to health or the environment.

Marketing and Use Directive

On the basis of this information, a hazard/risk assessment is made which may lead to the EC Commission to propose how to 'manage' the substance. This may take the form of an amendment to an existing Marketing and Use Directive. Alternatively it may include emission controls at any stage of production, processing, storage, transport, and disposal.

In addition to such measures, some Member States may take unilateral action to restrict a substance. This must, however, normally be done by notifying the Commission. We have represented these events in Figure 13.1.

The formulation and agreement of controls on a particular chemical, usually in the form of a Council Directive is subsequently implemented in Member States within the framework of national legislation.

There are three major aspects to controlling hazardous substances. These are:

- the protection of workers;
- the protection of consumers;
- the protection of the environment.

Thus EC-Directives controlling hazardous substances are implemented through the national legislation dealing with these three issues. We have represented this situation in Figure 13.2.

The mechanisms for implementing EC Directives controlling hazardous substances differs in different Member States. In the UK, for example, the implementation is achieved through Statutory Instruments of Acts of Parliament. Thus, the protection of workers is achieved through the implementation of The Health and Safety at Work Act 1974, consumers by The Consumers Protection ACt 1987 and the environment by The Environmental Protection Act 1990.

A general comment on the biotechnological production of chemicals other than medicines and food

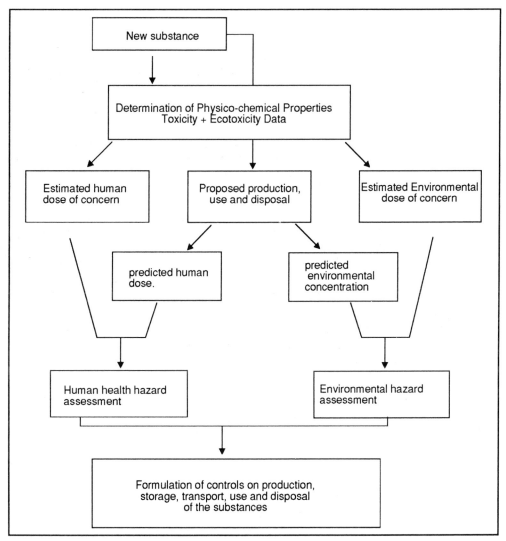

Figure 13.3 An overview of the hazard assessment of new chemicals and its interaction with the formulation of controls on its production, transport, use and disposal.

- its ecotoxicity.

The physio-chemical properties and toxicity data for the substances are determined for the substances by conventional laboratory-based procedures. These studies are conducted using Good Laboratory Practices (GLP - see Chapter 4). Ecotoxicology is a relatively new science. From the physico-chemical/toxicity data, an environmental concentration 'of concern' may be determined. From the proposed use of the substance, likely environmental concentrations may be calculated. This, of course, has to take into account the 'environmental fate' of the substance (for example, is it biodegradable? does it physically absorb onto solid matrices such as soil? what is its water solubility?) These data are used to make a hazard assessment and this in turn may be used to formulate controls concerning the production, use and disposal of the substance.

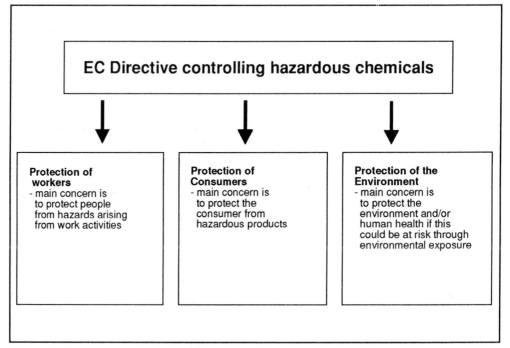

Figure 13.2 Overview of the implementation of EC-Directives controlling hazardous substances by Member States.

We have presented a generalised overview of the hazard assessment of new chemicals in Figure 13.3. It should be realised that the process is a non-linear one and many activities may be carried out in parallel.

A general comment on the biotechnological production of chemicals other than medicines and food

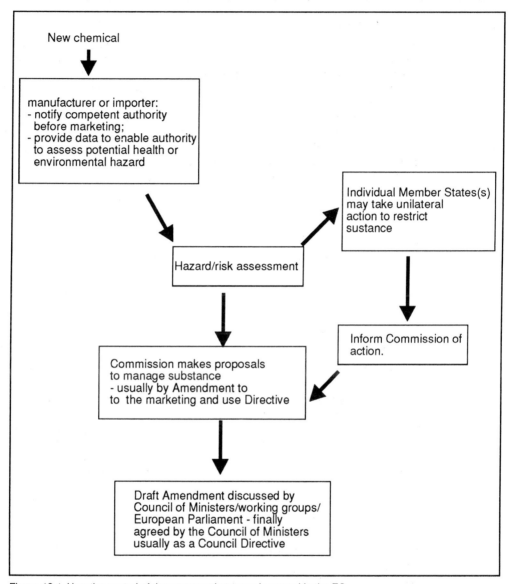

Figure 13.1 How the control of dangerous substances is agreed in the EC.

Although the implementation of these EC Directives differs in detail in each Member State, the principles that are aped are common throughout the Community. It is, of course, essential that you are familiar with the appropriate legislation within your region before introducing or using hazardous chemicals.

A key aspect of the control of hazardous substances is the hazard (risk) assessment process (see Figure 13.1). There are two main concerns, the effects of the substances on human health and the effects on the environment. In producing a new substance, the key properties that must be determined are:

• its physico-chemical effects;

• its toxicity;

Index

A

abridged applications, 245
access, 125
accommodation, 136
ACDP hazard groups, 89
ACRE, 123
addition of materials
    to a closed system, 121
additives, 302
Addresses of the offices, 19
adequate space, 265
adoption of written procedures, 84
advertising, 309
Advisory Committee on Dangerous Pathogens, 125
Advisory Committee on Genetic Modification (ACGM), 114, 123
Advisory Committee on Releases to the Environment (ACRE), 123
advocates-general, 11
aerosol, 95
aerosol dispensers, 15
aerosol generation, 95
African, Caribbean and Pacific
    EC Council ACR-ECC, 12
agricultural and agro-industrial research, 14
agricultural animals, 90
agriculture, 12 , 14
Agriculture and Agro-Industrial Research, 16
agriculutral levies, 13
air contamination, 95
Air Framework Directive, 62
Air lock, 278
air monitoring methods, 95
air quality standards, 15
Alpha decay, 190
alpha particle, 190 , 196
alpha radiation, 190
Alpha-particle emission, 191
amendments ,40
AMFEP, 300
amino acids, 300
Amoco Cadiz oil-spill disaster, 15
amphotrophism, 104
anaesthesia, 136
Ancillary areas, 285
Ancillary Community Bodies ,4 , 12
aneuploid stage, 102
animal cell cultures,102

animal cell lines, 102
animal experimentation, 52
animal health and welfare,15
animal tests, 271
animal varieties, 221
animals
    infected, 86
Animals (Scientific Procedures) Act 1986, 60 , 137
animals,
    germ free, 104
animals,
    use of, 136 , 137
antibiotics, 102
antibodies, 272
antigen-specific hybridomas, 103
antioxidants, 302
Apparatus, 72
apparatus and equipment,
    its maintenance and calibration, 264
apparatus and reagents, 74
appeal, 216
Appeal Board of the EPO, 219
Appeals Board within the Patent Office, 213
Appellate Court, 216
ARBO law, 115
Archive Facilities, 72
Association of Microbial Food Enzyme Producer, 300
Atom, 196
Atomic mass number,196
Atomic number, 196
attenuated vaccine viruses, 89
authorisation, 120, 131 , 132
authorisation procedure, for foods, 301
autoclave, 93 , 100

B

b+ emission, 191
b- emission, 191
baby hamster kidney cells, 102
bacteria, 102 , 270 , 272
bacterial
    aerosols, 96
bacterial fragments, 272
bacteriological contamination, 269
basic microbiological techniques (BMT), 91
batch number, 278
BATNEEC, 62
Becquerel (Bq), 194
bees, 90
Best Practicable Environmental Option (BPEO), 62
beta decay, 190
Beta particle,196